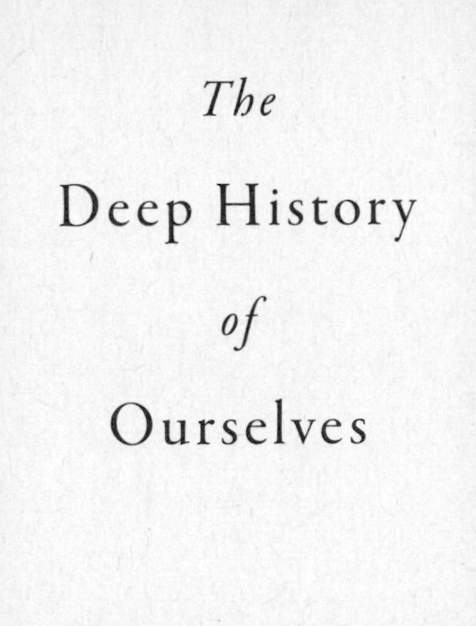

*The*

# Deep History

*of*

# Ourselves

Also by Joseph LeDoux

*The Emotional Brain*

*Synaptic Self*

*Anxious*

# *The* Deep History *of* Ourselves

## The Four-Billion-Year Story of How We Got Conscious Brains

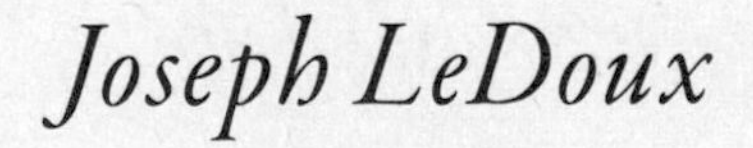


Illustrations by
Caio da Silva Sorrentino

VIKING

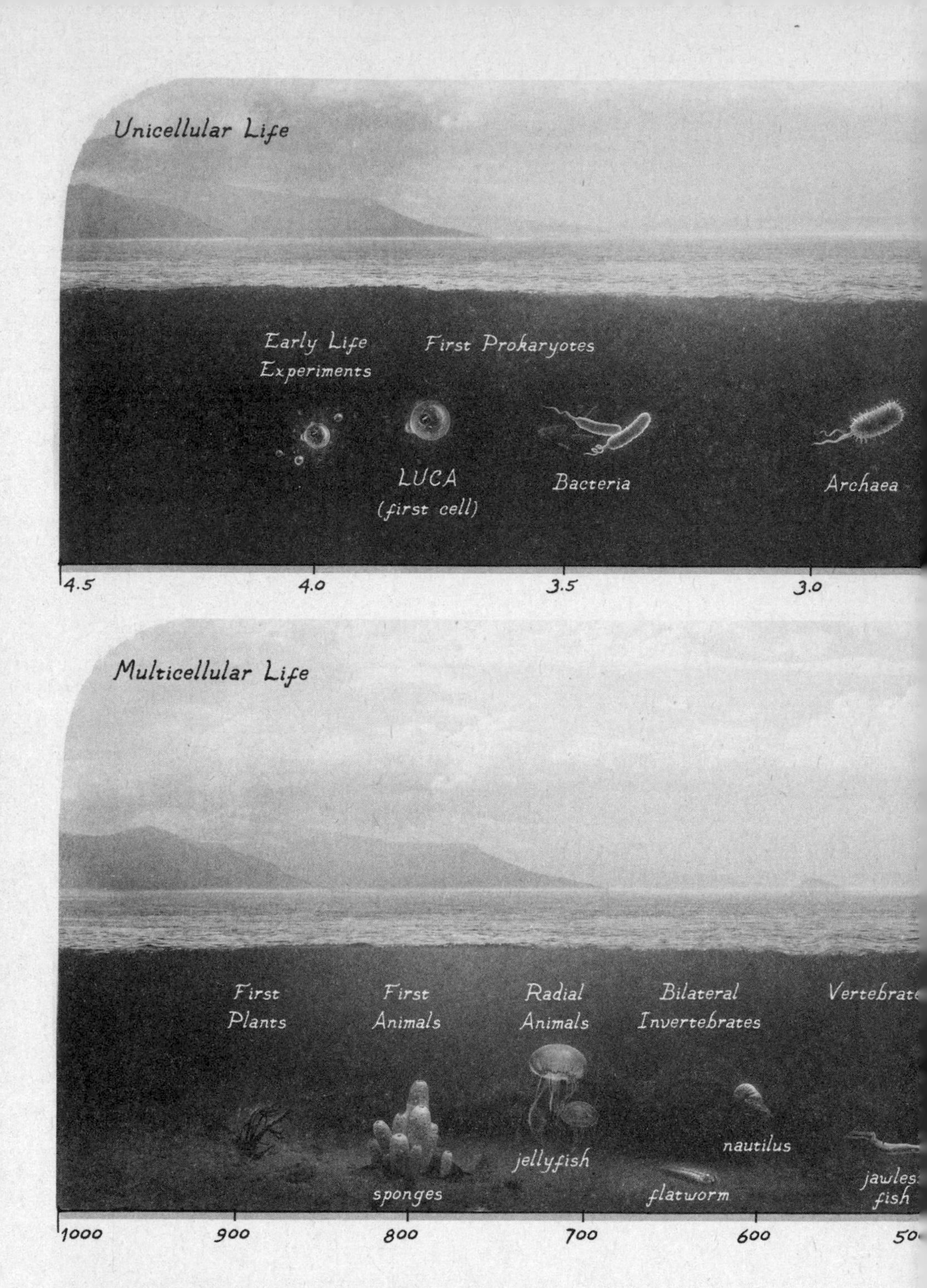


**Frontispiece:** ***Some Key Events in the History of Life***
*(for more details on the Timeline of Life see the table in the appendix)*

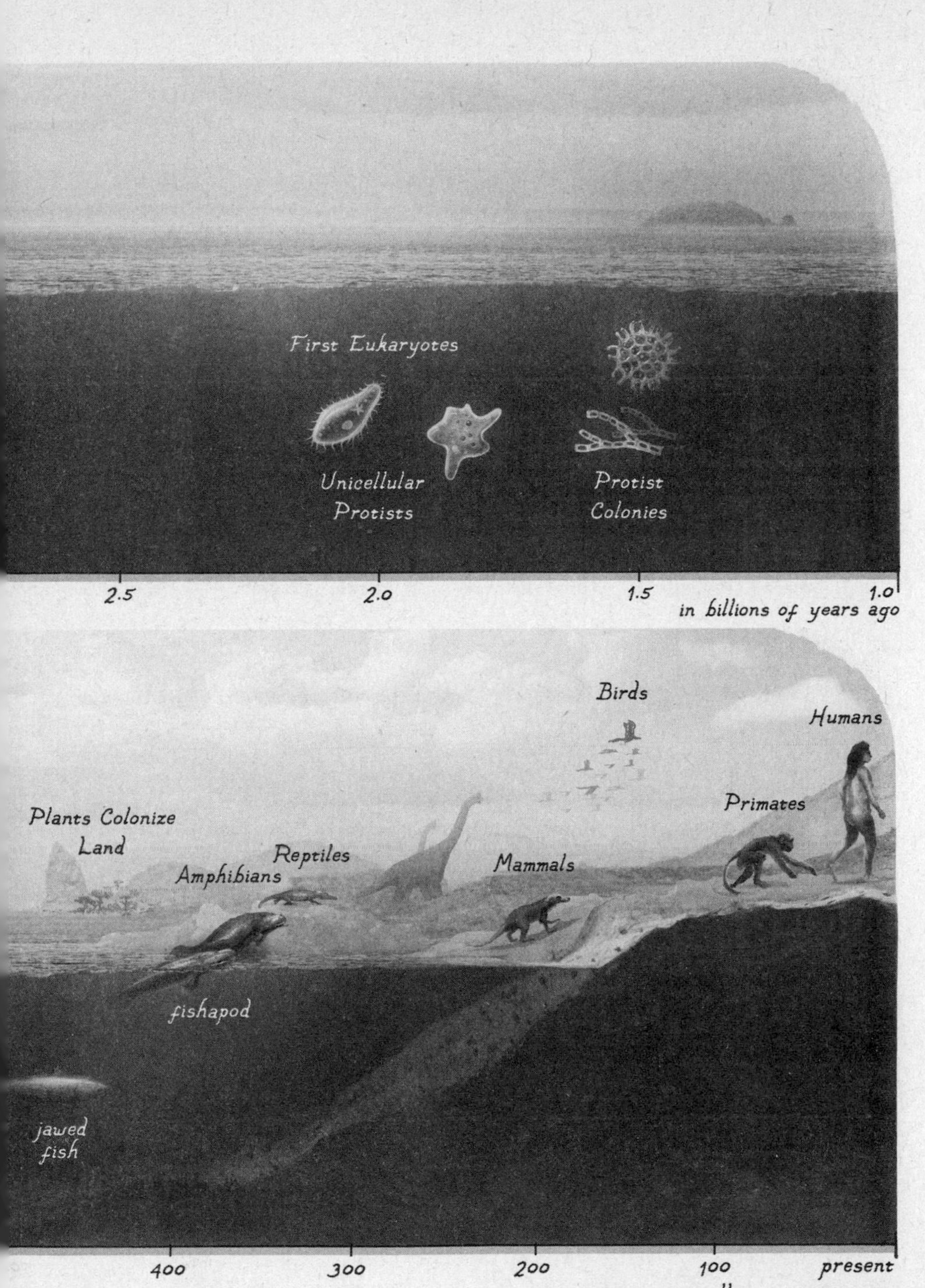
First Eukaryotes
Unicellular Protists
Protist Colonies
2.5
2.0
1.5
1.0
in billions of years ago
Birds
Humans
Primates
Plants Colonize Land
Reptiles
Amphibians
Mammals
fishapod
jawed fish
400
300
200
100
present
in millions of years ago

VIKING
An imprint of Penguin Random House LLC
penguinrandomhouse.com



LIBRARY OF CONGRESS CATALOGING-IN-PUBLICATION DATA
Names: LeDoux, Joseph E., author. | Sorrentino, Caio, illustrator.
Title: The deep history of ourselves : the four-billion-year story of how we got conscious brains / Joseph LeDoux ; illustrations by Caio Sorrentino.
Description: New York City : Viking, [2019] |
Includes bibliographical references and index. |
Identifiers: LCCN 2019011172 (print) | LCCN 2019012921 (ebook) |
ISBN 9780735223844 (ebook) | ISBN 9780735223837 (hardcover)
Subjects: LCSH: Consciousness. | Brain—Evolution. |
Nervous system—Evolution.
Classification: LCC QP411 (ebook) | LCC QP411 .L43 2019 (print) |
DDC 612.8/23—dc23
LC record available at https://lccn.loc.gov/2019011172

Printed in the United States of America
3 5 7 9 10 8 6 4 2

Set in Adobe Garamond Pro
Designed by Cassandra Garruzzo

*To Mike Gazzaniga,*
*my mentor and friend*

# CONTENTS

## PART 4: The Transition to Complexity



## PART 5: . . . And Then Animals Invented Neurons



## PART 6: Metazoan Bread Crumbs in the Oceans



## PART 7: The Vertebrates Arrive



## PART 8: Ladders and Trees in the Vertebrate Brain

## Part 9: The Beginning of Cognition



## Part 10: Surviving (and Thriving) by Thinking



## Part 11: Cognitive Hardware



## Part 12: Subjectivity



## Part 13: Consciousness Through the Looking Glass of Memory



## Part 14: The Shallows

# PREFACE

*The Deep History of Ourselves* is about the evolution of behavior. Not just the behavior of humans or other mammals or even other animals. It is about how behavior began as soon as living entities, organisms, came into being. These single-cell microbes, ancestors of the bacteria we share the planet with, had to do many of the same things we do to survive—avoid harm, obtain nutrients, maintain fluids and temperature, and reproduce. *Deep History* follows how subsequent organisms have used behavior to meet these same general survival requirements in their lives. But similarities only make sense in terms of differences, and a key goal of the book is to provide an account of the things most different about us: language, culture, our capacities of thinking and reasoning, and our ability to reflect upon who we are. These are new, but have deep roots that extend to the beginning of life.

Shortly before starting *Deep History,* I read E. O. Wilson's *The Meaning of Human Existence.* I was enthralled with the very compact "single idea" nature of his chapters, and decided to do the same. The chapters of *Deep History* are thus formulated as very brief "thoughts" or "meditations," stand-alone vignettes that cover a limited topic. I aimed to keep them to about fifteen hundred to two thousand words, and succeeded in most cases. I also wanted to follow Wilson's lead and write a short book, but did not do so well on that front.

The chapters are grouped thematically, so that if you want to learn specifically about how life began, or bacterial behavior, or how sexual reproduction emerged, or how life progressed from single to multicellular existence, or how nervous systems evolved, or about the key role of sponges and jellyfish in human evolution, or how cognition or emotion evolved, or what we know about consciousness and the brain, you can just read the relevant sections that interest you. But for those who read from start to finish, *Deep History* will take you on an ascent of the tree of life in such a way as to connect the survival capacities of ancient microbes to our own unique capacities to survive and thrive by thinking and feeling, to contemplate our personal past and future, and the future of our kind.

This is my fourth single-authored book. I learned something important in writing the first one. The best way to figure out what you really know—and don't know—about your field is to write about it. *Deep History,* though, was a little different. I knew from day one I had a lot of research to do if I was going to write about the history of life. Much of the first part of the book is thus written with me serving more as a scientific journalist than as an expert. Consequently, I sought the help of experts when I felt I was completely out of my depth (hopefully, I did enough of this). When I got to the parts I "knew" about, I got to have my usual "aha" experiences about how much I had to figure out, and then also reached out to colleagues.

Thanks very much to all of you who consulted, including Tyler Volk (prebiotic chemistry and early life), Nick Lane (origins of life), Karl Niklas (origins of multicellular life and the role of export- and alignment-of-fitness), Sarah Barfield (germline segregation), Ralph Greenspan and Takeo Katsuki (jellyfish behavior), Iñaki Ruiz-Trillo (protozoan ancestors of multicellular organisms), Linda Holland (early origins of bilateral animals; divergence of protostomes and deuterostomes; divergence of chordates from other deuterostomes, and vertebrates from other chordates), Maja Adamska (sponge physiology and behavior), Sten Grillner (early vertebrate nervous systems), Eric Nestler (epigenetics and behavior), Betsy Murray (evolution of perception and memory systems), Charan Ranganath (perception and memory), Cecilia Heyes and Thomas Suddendorf (ruling out nonconscious explanations before claiming consciousness in humans or animals), Nathaniel Daw (cognitive deliberation), Marian Dawkins (anthropomorphism), Liz Romanski, Helen Barbas, Roozbeh Kiani, and Todd Preuss (prefrontal cortex), Hakwan Lau and Steve Fleming (metacognition and consciousness), Karl Friston (predictive coding), Richard Brown (philosophy of mind), David Rosenthal (higher-order thought theory of consciousness), and Christophe Menant (self, consciousness, and evil).

I wanted *Deep History* to have a visual as well as a verbal narrative. When I mentioned this to an artist friend, Heide Fasnacht, she told me that she had a talented student who might be worth considering. Caio da Silva Sorrentino's work impressed me, and once he generated some samples, based on preliminary text that I had written, I was sold. Caio developed the perfect visual concept for the art, one reminiscent of biological illustrations from the late nineteenth century, including hand-drawn lettering. Caio is indeed a talented illustrator, and I have enjoyed working with him. *Deep History* greatly benefited from his vision.

When writing my last book, *Anxious,* my wife, Nancy Princenthal, was busy with her book on the artist Agnes Martin, which went on to win the PEN America Award for Biography. We basically started and finished around the same time, and our marriage survived the horse race. It also survived our in-tandem efforts being completed now—*Deep History* for me and a book on sexual violence in the art of the 1970s for her. I could not have gotten mine done quite the same without her moral support and also expert editing of parts I was struggling with. Our son, Milo LeDoux, a classicist turned capital-markets attorney, contributed key millennial insights from time to time over dinner. For example, when I mentioned something about human behavior often being nonconsciously controlled, he quipped, "It's sort of like being behind the wheel of a Tesla." He gets a footnote or two in the book.

Rick Kot, my editor at Viking and friend in life, has shepherded me through my last three books. He has helped shape each one conceptually, and tidied up my text when I would get too "weedsy." I couldn't ask for a better publishing partner than Viking. The entire staff has been fabulous on this and my other books with them. On this one, Norma Barksdale, Katie Hurley, Claire Vaccaro, and Cassie Garruzzo were especially helpful.

Thanks also to my agent, Katinka Matson at Brockman Inc. She has held my hand through my four books, sharpening the proposals to make them appealing to publishers and providing helpful guidance along the way.

New York University has been a wonderful academic home to me since 1989. The university, and my home department, the Center for Neural Science, have supported me every step along the way, and made it possible for me to be both a scientist and writer, and, in some ways, even helped me be a musician.

William Chang has run my office at NYU for almost two decades. He has been a key player behind the scenes in keeping my text and illustrations in order, and in so many other ways. Claudia Farb and Mian Hou, two other multidecade staff members, have also helped in ways that defy words. I also thank all the undergraduate and graduate students and postdoctoral researchers and visiting scholars and other scientists who have been research collaborators over the years.

Finally, I thank past and present members of my band the Amygdaloids for their friendship, musical collaboration, and intellectual stimulation, including Tyler Volk, Daniella Schiller, Nina Curly, Gerald McCollum, Amanda Thorpe, and Colin Dempsey. We are "rock-it" scientists.

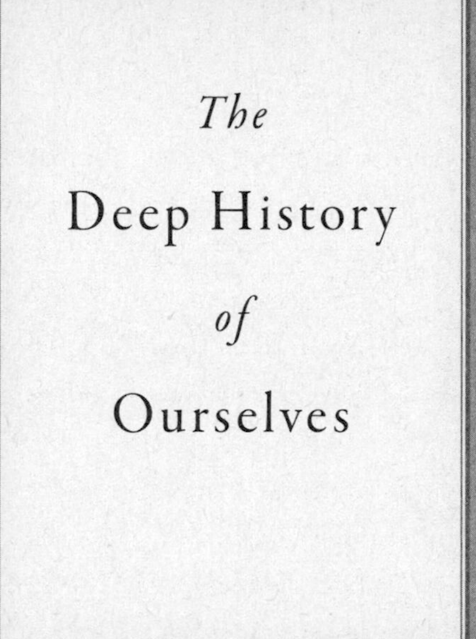

*The*

Deep History

*of*

Ourselves

*Prologue*

# WHY ON EARTH . . . ?

When I mentioned to a friend that I was writing a book about the history of life, she asked, Why on earth are *you* doing that? She knew that I had spent most of my scientific career studying circuits in the rat brain that underlie behavioral reactions to danger, with an eye on how that information could help us understand at least some aspects of human emotions, especially fear and anxiety.

Part of the answer to my friend's question is that if we really want to understand human nature, we have to understand its evolutionary history. As the biologist Theodosius Dobzhansky once said, "Nothing in biology makes sense except in the light of evolution," and that includes behavior.

That behavior and evolution are interrelated is hardly a novel idea. Darwin emphasized it, as did pioneering ethologists such as Niko Tinbergen and Konrad Lorenz. The behaviorists who dominated psychology in the first half of the twentieth century paid little attention to evolution, but contemporary psychologists and neuroscientists accept it as a key factor.

Most efforts to understand the evolution of behavior, especially in neuroscience, typically focus on the relation between closely related groups, such as humans and other mammals. There are obvious reasons for doing so. For example, since the brain controls behavior, studies of how brains evolved in such groups help explain the evolution of their respective behavioral repertoire, and also ours. But there are also good reasons to look deeper. Research comparing mammals (often rodents) and invertebrates (such as flies and worms) shows the connection between these, and is revealing how memory works in us. But in this book I have opted to dive even deeper, in fact, *very* deep—all the way back to the beginning of life, and even earlier, to the so-called prebiotic chemical conditions of the Earth that made biology, and hence life, possible.

I had always been casually interested in the evolution of the brain and behavior, but never pursued the topic with much vigor. Then, in 2009, I spent some time at Cambridge on sabbatical and became friendly with Seth Grant, a neurobiologist whom I first met while he was a postdoc uptown working in Nobel Laureate Eric Kandel's lab at Columbia. While there, he began researching the evolution of genes involved in synaptic plasticity to better understand the biological mechanisms of learning and memory, and was continuing this line of work at Cambridge.

Seth found parallels in plasticity-related genes between rodents and sea slugs, suggesting that they may each have inherited the ability to learn from a common ancestor that lived many hundreds of millions of years ago. But even more interesting, some of the same genes exist in single-cell protozoa. That's relevant, since animals and current-day protozoa share a common protozoan ancestor that lived over a billion years ago. Some of the learning-related genes in our nervous system may therefore come to us via such microbial ancestors.

If you know anything about protozoa, you may be scratching your head regarding these findings. Most people, if they think about it at all, think of behavior, and especially learned behavior, as a product of a nervous system. But protozoa, being single-cell organisms, don't have nervous systems, since that would require special cells—neurons—and they only possess one all-purpose cell. Yet they have a robust behavioral life—they swim away from harmful chemicals and toward useful ones—and they even use past experience to guide their present responses, suggesting that they have the ability to learn and remember. The logical conclusion is that behavior, learning, and memory don't actually require a nervous system.

This was eye-opening to me, so I did a little research to see what was known about the behavioral capacities of these single-cell organisms. I found accounts not only of their swimming away from danger and toward nutrients, but also of moving toward or away from chemicals or sunlight to balance fluids or regulate temperature inside the cell relative to its environment. Protozoa even engage in mating behavior—sex—to reproduce their kind.

Protozoa are relatively recent single-cell organisms, having appeared about two billion years ago, when they evolved from another familiar single-cell creature—bacteria, which are the oldest living organisms, having emerged about 3.5 billion years ago. Bacteria exhibit many of the same kinds of behaviors that protozoa do, but did so first. They approach and avoid useful and harmful things in their world, and may even learn from experience what is useful and harmful in their world. They don't, however, reproduce sexually; they simply divide in half. Sex is the behavioral claim to fame of eukaryotes, which evolved from bacteria, and which include protozoa and animals.

When animals engage in defensive, energy management, fluid balance, and reproductive behaviors by freezing/fleeing, eating, drinking, and mating, scientists and laypeople alike often describe these activities as an expression of underlying psychological states—consciously felt experiences such as fear, hunger, thirst, and sexual pleasure. In doing so we effectively project our own experiences onto other organisms. Given how ancient these behaviors are, and how they arose long before nervous systems, we should probably be more judicious in making such attributions based on our mental states.

I will argue in this book that such survival behaviors have deep roots that date back to the beginnings of life. Animals later evolved neurons and circuits to make these behaviors more efficient and effective. But all organisms, whether composed of only one or of billions of cells, engage in these kinds of survival activities in the process of staying alive and well.

Because humans consciously experience feelings when we engage in our own survival behaviors, we intuit that these feelings and the behaviors must be intrinsically related—that the feelings are the causes of the behaviors. And because other animals close to us (other mammals, for example) behave like us in survival situations, and the circuits that control these behaviors are similar in us and them, their survival behaviors must be driven by their feelings.

But I will present evidence that turns this logic on its head. I will show you that there is indeed good evidence that the same brain systems control survival behaviors in humans and other mammals, but that these are not the systems that are responsible for conscious feelings we experience when we engage in such behaviors. The behaviors and feelings occur simultaneously, not because the feelings drive the behavior, but because their respective systems are responsive to the same stimuli.

Survival behaviors thus have very old roots that make them universal. But the kinds of experiences humans call conscious feelings—that is, emotions—I propose are a much more recent development, possibly emerging via evolutionary changes in the human brain a mere few million years ago that brought language, culture, and self-awareness to our species. This idea is likely to be controversial to some, perhaps many, since it seems to deprive animals of conscious experiences. But I hope that even those who are skeptical about this view of animal consciousness will hear me out.

I don't, in fact, deny that animals have conscious experiences. My point, instead, is that any conscious experiences they have are likely to be very different from ours, as every species has somewhat different brains. And given that brain circuits with unique properties not found in other animals, not even other primates, are emerging as potentially important for the kinds of conscious experiences we have, we should be cautious in attributing humanlike conscious experiences to other creatures. But to say that they don't have the kinds of experiences we have does not mean they have no experiences. For example, that they do not suffer the way we do does not mean that they do not suffer. While it is extremely difficult to scientifically assess consciousness in other animals, toward the end of the book I will speculate about what kinds of conscious experiences nonhuman primates and nonprimate mammals might possibly be capable of, given the kind of brain each possesses.

Recognition that the circuits that control our survival behaviors are different from those that assemble our emotions and other conscious experiences allows us to see our connection to the deep history of life in a new way. Like all species, we are similar to the species from which we evolved, but are also, by definition, different. And to fully appreciate our differences, we have to be as precise as possible about both our similarities and differences, using science, not intuition, to draw the conclusions.

Though I had been thinking about the ancient biology of survival since 2009, my first public foray into this topic began with a 2012 article entitled "Rethinking the Emotional Brain." I have continued this line of thought in other writings and lectures ever since. *Deep History* consolidates and extends the nascent ideas I had been mulling over into an overarching vision of how we came to be who we are, taking the argument all the way from the beginning of life in primordial microbes to the emergence of our ability to consciously know about the existence of our selves and our thoughts, memories, and emotions.

# Part 1

# Our Place in Nature

*Chapter 1*

# DEEP ROOTS

"We are our brains." This statement is viewed as incontrovertible by some and preposterous by others. What is clear is that the essence of who we each are depends on our brains. Brains enable us to think, to feel joy and sorrow, to communicate through speech, to reflect on the moments of our lives, and to anticipate, plan for, and worry about our imagined futures.

The evolutionary history of the human brain (figure 1.1) is usually recounted in terms of animals relatively close to us, especially mammals and other vertebrates. We hear a lot about how humans have neural capacities that allow more sophisticated behavioral and cognitive features than those

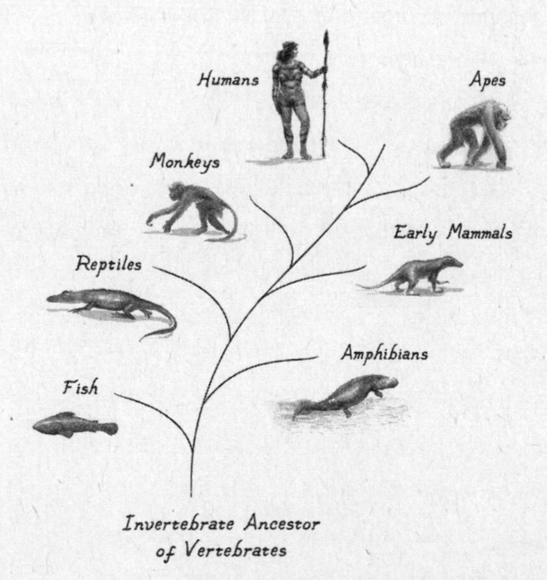


**Figure 1.1:** ***Evolution of Humans from Vertebrate Ancestors***

of our primate ancestors who possessed brains that diverged from those of other mammals. We also hear that mammals and their brains evolved from reptilian ancestors, and that reptiles and amphibians evolved from fish, which in turn evolved from invertebrate ancestors.

Insights about the psychological functions of other vertebrates clearly provide insights into how our brains have come to be what they are, and how they contribute to our own psychological essence, the core of who we are, including those things we like about ourselves, but also characteristics we'd rather not have. But I will argue that this common strategy can take us only so far. It is like trying to understand the history of digital computers by starting with the first devices that were cosmetically similar to today's computers—Commodore, Apple, and IBM "personal computers" of the late 1970s. In fact, the essence of digital computing had long been figured out by the time these devices were introduced. So if you want to know how present-day computers work, you can learn a lot by studying them and their immediate forerunners. But if you really want to know how they came to be what they are, and why they do what they do, you have to know their deep history—their evolution from analog ancestors, including nonelectronic ones (the abacus, for example).

Similarly, to truly understand the complex psychological functions of our brains, we have to take the long—in fact, the very long—view of history. Like the roots of a tree, the deep roots of our brain and its cognitive and behavioral capacities are not visible; we must dig down to unearth and understand them. And we must dig past other mammals, other vertebrates, and even beyond the invertebrate ancestors of vertebrates. We need to go as deep as ancient unicellular microorganisms, the original life-forms on Earth.*

Why do we have to look so far into the past to understand the origins of the human brain and its functions? Unicellular organisms like bacteria don't have a nervous system, much less a brain. Why can't we just focus on

---

* Other recent books that delve into ancient biology in relation to the human mind include Daniel Dennett's *From Bacteria to Bach and Back Again*, Antonio Damasio's *The Strange Order of Things*, and Arthur Reber's *The First Minds*, each of which has a different perspective and reaches different conclusions than I do.

animals with brains, or at least nervous systems, if our goal is to understand the human brain? A key argument of this book, as discussed in the prologue, is that core survival requirements were solved by the first successful living organisms billions of years ago, and that they passed their solution on to every organism that followed.

As organisms evolved to consist of many cells, the problem of controlling their behavioral activities in the quest to survive became more complex, requiring the coordination of the activities of cells distributed in different parts of the body. This led first to simple diffuse nervous systems in creatures like hydra and jellyfish, and eventually to nervous systems with central control units, or brains.

Cellular features that enabled early primitive organisms to survive and reproduce were thus carried forward and underlie the history of life as we know it (see frontispiece). But my point is not to trace the biological interconnectedness of organisms on Earth—a project that has been undertaken many times before—but rather to show that the roots of behaviors that humans routinely call upon in day-to-day life are more ancient than we generally acknowledge.

Behavior is not, as we commonly suppose, primarily a tool of the mind. Of course, human behavior can reflect the intentions, desires, and fears of the conscious mind. But when we go deep into the history of behavior, we can't help but conclude that it is first and foremost a tool of survival, whether in single cells or more complex organisms that have conscious control over some of their actions. The connection of behavior to mental life is, like mental life itself, an evolutionary afterthought.

To truly appreciate how our brains make it possible to be who we are, we need to understand survival strategies that were built into ancient single-cell organisms, maintained in primitive multicellular life-forms, taken over by specialized cells called neurons when nervous systems developed in early invertebrates, retained in nervous systems of invertebrate ancestors of vertebrates, and subsequently used by humans and all other animals in daily life, regardless of how simple or complex their body is.

Only by examining the natural history of life on Earth can we appreciate the unique features that were added to organisms incrementally over the course of evolution and that resulted in the kind of brain we have and the functions embodied within it. This does not mean that human behavior can be fully understood in terms of ancient survival responses. It means that we need to be clearer about which aspects of human behavior *are* related to processes inherited from various other organisms, so that we can better understand those that are not.

*Chapter 2*

# THE TREE OF LIFE

Until the second half of the nineteenth century, the interrelation of living organisms was generally envisioned as a progression that placed humans above all other earthly creatures. Aristotle's ranking, called the "scale of nature" (*scala naturae*) or "ladder of life," was based on the complexity of observable features of organisms. Humans, being the most complex, were at the top. The presence of blood divided what we now called vertebrates from invertebrates, and within invertebrates those with a shell (e.g., clams, mussels, and oysters) were considered as a separate class, intermediate between animals and plants. Christian theologians in the Middle Ages, building on the creation story of Genesis and Aristotle's scale, proposed a ranking based on "perfection," considered as closeness to God. In this "Great Chain of Being," humans, made in God's image, were the most perfect of all organisms on Earth. Particularly notable in this tradition is the notion that all life began more or less simultaneously, roughly six thousand years ago, when God populated the Garden of Eden, most famously with people, apples, and snakes; and once created, the distinct organisms continued to exist in the same form, unchanged over the millennia.

In the nineteenth century, a different view began to emerge through the writings of Alfred Russel Wallace and Charles Darwin. Darwin, though, captured the spotlight of history, and will be our focus here.

In his 1859 book, *On the Origin of Species,* Darwin proposed that present-day organisms evolved from earlier forms over long periods of time—in fact, much longer than the several thousands of years proposed in the biblical account. Building on his own observations of nature, and those of earlier philosophers and scientists, Darwin argued that the relation between organisms is more like a branching tree, a "tree of life," than the linear scale implied by metaphors based on steps or ladders. In Darwin's words: "The green and budding twigs may represent existing species; and those produced during former years may represent the long succession of extinct species."

At the base of the trunk lies what Darwin called the "primordial form," an organism from which all other life-forms evolved, in piecemeal fashion, with some adapting and surviving and others becoming extinct. Breaking with the Judeo-Christian tradition, which treated organisms as unique, unrelated creations, Darwin argued that all living organisms are connected to one another by way of this common ancestry.

While the tree of life is a more accurate scientific depiction of nature than the ladder or scale view, it does not always successfully circumvent the conclusion that humans have a special place in nature. For example, the late-nineteenth-century biologist Ernst Haeckel popularized Darwin's tree of life in several illustrated publications. One famous one, called the Pedigree of Man (figure 2.1), depicts a tree with branches, but has the major groups of animals stacked up on top of one another along the trunk, with humans occupying the topmost part of the tree. It is unlikely that Haeckel, a pioneering biologist who made numerous contributions in his own right, intended to imply that humans are the result of a single linear progression of life. Regardless of his intention, though, his diagram suggests to a naïve viewer that humans are the "highest" organisms on Earth, the end point of the ancient progression of life.

Mid-twentieth-century textbooks continued to place humans at the top of the tree of life. This was in part driven by a prominent theory of brain evolution that viewed the human brain as a mélange of the brains of our vertebrate ancestors, with a reptilian brain beneath an early mammalian brain, which in turn was below a later, and "higher," kind of mammalian brain possessed by primates, with the human brain being the pinnacle. The presupposition that we are the top of the evolutionary heap

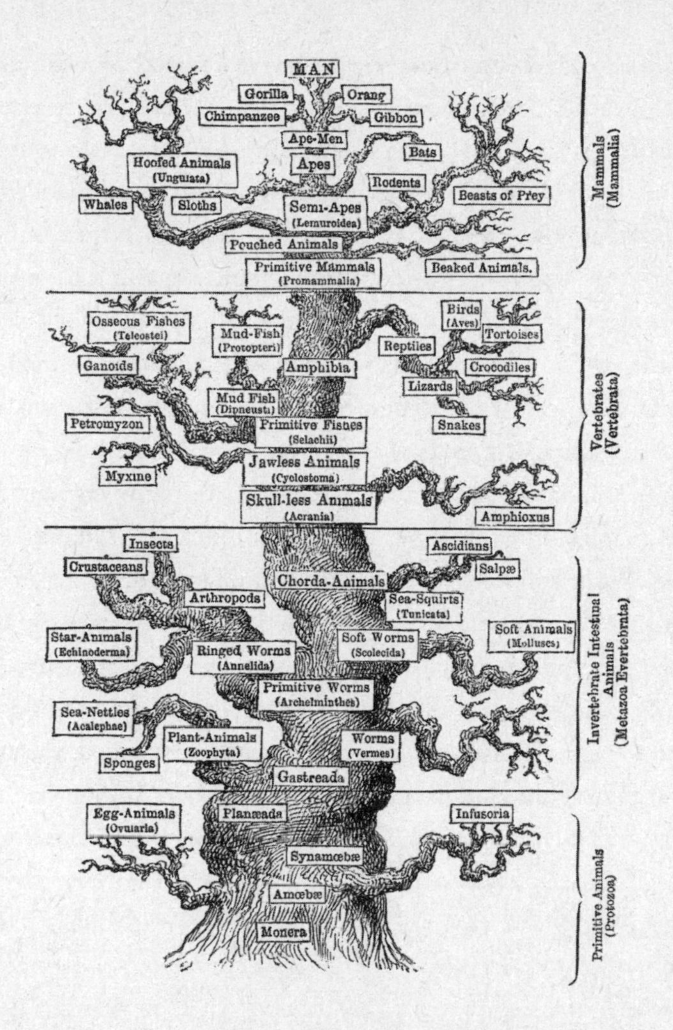


**Figure 2.1:** ***Haeckel's Pedigree of Man***

continues to influence views about brain evolution and function in some circles, as well as about the nature of the human mind and the realms of ethics and morality. It's very difficult for humans to abandon the idea that we are special, the purpose of life.

Human "exceptionalism" is for many people an unquestioned assumption. For the religious, it is a God-given fact; for humanists, it is a celebration of our special powers of thought and feeling. But just as our species emerged through change, we, too, are constantly changing. Each time a baby is conceived, it receives a unique genetic makeup that never existed before. What is considered special today may well turn out to be mundane in the organisms that diverge from us in the future; they will have their own traits that make them different from us, and thus special in their way.

In some sense, then, we are special, if by "special" we mean different. Using that criterion, every organism, being a distinct form of life, can be considered special. The challenge, though, is to pinpoint exactly how we differ from other species so that we neither anthropocentrically deny features deserved by other organisms, nor anthropomorphically attribute features to other organisms that they may not possess.

When we talk about our family history, our "family tree," we are referring to a set of tiny sprouts on twigs far, far out on the human twig of the tree of life. Just as your family history is of interest to you because it is your history, the evolutionary history of our species is our history, and of particular personal interest to us. But our species is only another twig on just another branch of the tree. It's a twig we care about, and a point from which to view the history of life, but not because we occupy a special place in the natural order of things.

*Chapter 3*

# KINGDOMS COME

The classification of the natural world has long been a preoccupation of scientists. The Greeks referred to this exercise as the "carving of nature at its joints." Aristotle carved the natural world into three categories—plants, animals, and humans. The difference between them, he explained, is that plants are limited to nutritional and reproductive capacities, while animals additionally have sensation and movement; and humans are distinct from mere animals in that they also had thought or reason. These capacities reflected three kinds of souls: plants have a vegetative soul; animals, which he called brutes, a sensitive (sensory) soul; and humans a rational soul.

The categories that result from such efforts are called kingdoms, the number of which has been the subject of much debate (table 3.1). In the sixteenth century, the famed taxonomist Carl Linnaeus revised Aristotle's scheme, dividing life into plants, animals (including humans), and minerals. The invention of the microscope in the seventeenth century led to the discovery of microbes, single-cell organisms invisible to the naked eye. Ernst Haeckel named these protists (meaning "the first") and added Protista to Plantae and Animalia to form a three-kingdom system of life. With the invention of the electron microscope in the twentieth century, it was discovered that some microbes have a cell nucleus, where DNA is sequestered, while in others there is no nucleus and DNA is distributed throughout the cell. Unicellular organisms with a nucleus (e.g. amoeba, paramecia, and algae) remained in Protista, while those lacking a nucleus (bacteria) were placed in the kingdom Monera, which was divided into two groups, Bacteria and Archaebacteria (commonly known as Archaea). Later, fungi (mushrooms and yeast) were added as another multicellular group, giving us the six-kingdom view, which is the most widely accepted way to classify nature today.* As we will see later, organisms without a nucleus (bacteria and archaea) are called prokaryotes, while those with cells containing a nucleus are called eukaryotes (protists, plants, fungi, and animals).

TABLE 3.1: Evolving Views of the Kingdoms of Life

| 2 KINGDOMS | 3 KINGDOMS | 4 KINGDOMS | 5 KINGDOMS | 6 KINGDOMS |
|---|---|---|---|---|
| Plants | Plants | Monera | Monera | Bacteria |
| Animals | Animals | Protista | Protista | Archaea |
| | Minerals | Plants | Plants | Protista |
| | | Animals | Fungi | Plants |
| | | | Animals | Fungi |
| | | | | Animals |

Modern biologists continue to use the branching-tree metaphor to represent relationships within and between kingdoms, but are careful to avoid the implication that humans are an end point in an evolutionary progression. In contemporary depictions of the tree of life, the six kingdoms are each represented as a major limb of the tree, with different groups of organisms within each depicted as branches of their kingdom's limb (figure 3.1).

Classification does not stop with kingdoms, as there are many different subgroups of organisms within them. Linnaeus sought to provide a rank-order system for organizing these groups into progressively more restricted categories. Building on Linnaeus, it is common to classify organisms within kingdoms into phylum, class, order, family, genus, and species. For our

* Some argue for seven- and eight-kingdom systems, but we can tell our story without venturing into this controversy.

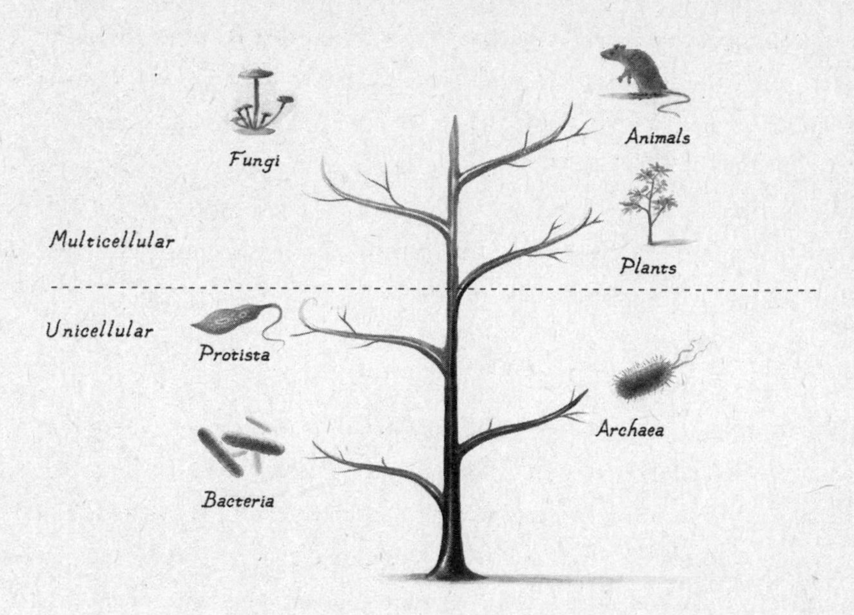


**Figure 3.1: *The Six Kingdoms Shown in a Tree of Life***

own species, the classification is: kingdom Animalia, phylum Chordata, subphylum Vertebrata, class Mammalia, order Primate, family Hominid, genus *Homo,* and species *sapiens.*

Classification schemes in no way represent a return to the notion that different kinds of organisms are unique de novo creations. They are instead very much in keeping with Darwin's suggestion that every organism alive now is, in one way or another, related to all present and past members of each of the kingdoms, because all forms of life arose from a primordial common ancestor.

*Chapter 4*

# COMMON ANCESTRY

There are two broad ways to trace evolutionary history. One is through the fossil remains of organisms in different layers or strata of rocks. By dating the rocks, the period in which the organism lived can be estimated. This method is limited by the fact that many organisms have soft tissues that do not fossilize well. The other approach is to use the tools of modern genetics to reconstruct history with much more precision, since an organism's genes are the record of its evolutionary history.

Scientists have been able to identify Darwin's primordial form by comparing the genes from different kingdoms of organisms and isolating common genes that date back to the beginning of life. As Nick Lane writes in his book *The Vital Question,* the ancestor of all life on Earth is a cell that arose between 4 billion and 3.8 billion years ago, about a half a billion years after the Earth was formed. This cell, he says, was quite sophisticated, and by virtue of having survived, passed its survival-relevant traits on to every subsequent cell that has lived since, including all the cells in our bodies. The primordial organism, the mother of all life, has a name, or a nickname: she's called LUCA, the last universal common ancestor of all of life, the base of the tree of life.*

LUCA was likely not the first instance of life. A cell is a complicated piece of biological machinery, and LUCA didn't suddenly just appear fully

* This account, some argue, is too neat—that LUCA was not a single cell or even a single kind of cell, but a collection of a number of cells. For example, Carl Woese, another leader in this area, has noted, "The ancestor cannot have been a particular organism, a single organismal lineage. It was communal, a loosely knit, diverse conglomeration of primitive cells that evolved as a unit, and it eventually developed to a stage where it broke into several distinct communities." These communities, he says, became the three primary lines of descent—bacteria, archaea, and eukaryotes. We will stick with the idea of LUCA here, while recognizing the possibility that LUCA may be a metaphor for a community of early cells rather than an actual cell or cell type.

formed. RNA, DNA, and proteins formed primitive forms of life, called protocells, in the early waters of the world. From these biological fits and starts, LUCA eventually emerged. While LUCA herself no long exists, her first children—Bacteria and Archaea—do, and they form an essential link in the evolution of all subsequent life, including animals like us.

The key to understanding evolutionary history is thus the notion of common ancestry. A group of organisms with a common ancestor is called a clade. Evolutionary biologists often use cladograms to show common ancestry based on fossil or genetic evidence. The six kingdoms of life that evolved from LUCA are shown in the cladogram in figure 4.1. The branch points from the long diagonal line in a cladogram represent common ancestors, which are often not living organisms—99 percent of all species that have ever lived are extinct (common ancestor branch points are shown as dots in this first cladogram). The ends of the lines extending out from branch points represent living organisms.*

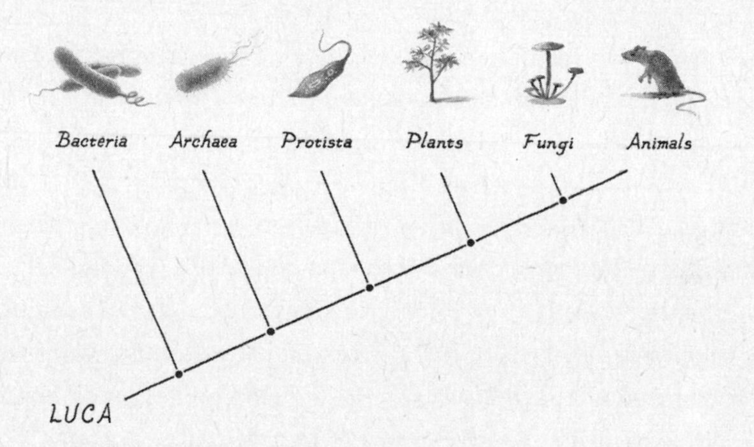


**Figure 4.1:** ***The Six Kingdoms of Life Shown in a Cladogram***

This is a good place to dispel the incorrect idea that that humans evolved from monkeys. The famous "Scopes Monkey Trial" that pitted

* When reading the cladogram in figure 4.1 you may be tempted to conclude that fungi and animals evolved from plants. This would not be the correct interpretation. Plants, fungi, and animals each have a unique protist ancestor that, in turn, evolved from an even more ancient common protist ancestor.

evolution against religion was based in part on this false assumption. Humans are hominids, and our closest nonhominid primate relatives are chimpanzees. So, if anything, it should have been the "Chimp Trial." But the fact is, neither did we evolve from chimps. Instead, early hominids shared a common ancestor with chimps (figure 4.2).

Similarly, although rats and humans are both mammals, and thus

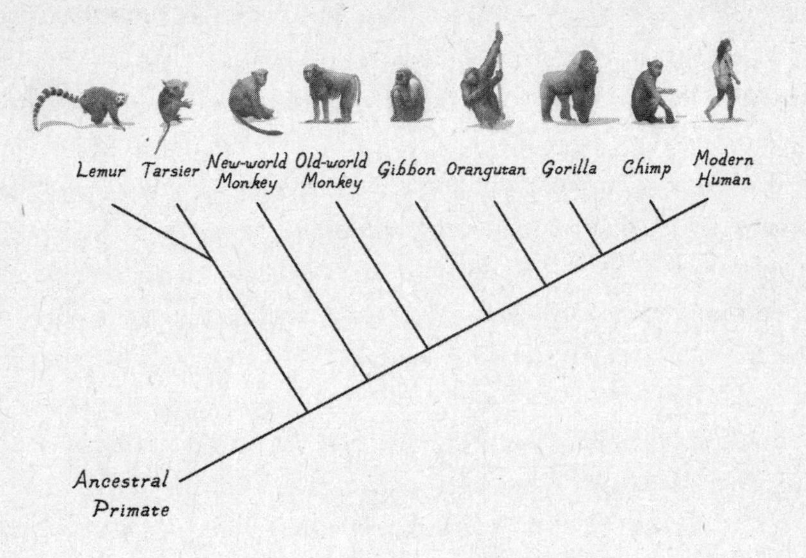


**Figure 4.2:** ***Cladogram of Primate Evolution***

share certain features, such as body hair and nursing of young, we have no rats in our past. Instead, rats and primates share a common mammalian ancestor from which these features were inherited. This is why we can learn some important things about the human brain by studying that of rodents. In doing so, though, we have to be careful not to assume or imply that a human is simply a more advanced rat, or even a more advanced chimp, as these animals are in different clades.

Darwin proposed that the way that clades emerge from common ancestors is through natural selection. His theory assumed that traits vary in a population, and those traits that are useful to survival increase in frequency because those individuals who have them are more likely to survive and reproduce. In other words, the useful traits are selected and stabilized.

The later discovery that genes and chromosomes are the mechanisms of inheritance provided strong support for the theory.

Natural selection would seem to continuously drive organisms toward a better and better fit to their environment. And while that sounds good, as the old saying goes, you can have too much of a good thing. If a shortage of food supplies or the arrival of new predators necessitates a move, a group's being extremely well adapted to their current environment could be a disadvantage, as it may have shed traits that could have given it more flexibility to adapt to new environments where other traits would be useful.

Besides selection and migration, other factors are important in evolution. Genetic mutations are a key source of variability. Mutations are best known for their deleterious effects, such as in diseases, but they can also result in traits that are beneficial to survival and can thus come to be stabilized in the population as well. Genetic drift—random changes in the frequency of certain genes in a population—also affects evolution. For example, imagine a hurricane that causes the death of many inhabitants of an isolated island. If many of those who died had genes for a certain trait, the frequency of that trait could decrease in the future population. As different traits become more prominent, a new clade can come to exist.

An important distinction should be made at this point between features that are derived from a common ancestor and those that are similar in different species for reasons other than common ancestry. The vertebral column (backbone or spine) possessed by all vertebrates is a feature derived from the common ancestor of all vertebrates, and the possession of body hair by mammals was passed on from the furry common ancestor of all mammals. The opposable thumb of primates and pandas, however, is a feature that was not present in the common mammalian ancestor of these animals and thus evolved separately to solve a similar problem. Also, there is no winged common ancestor that links bats to birds—bat wings actually evolved as modifications of the forearm of their nonwinged mammalian ancestors. Differences based on common ancestry are said to be homologous, while those that are not based on common ancestry are merely analogous.

*Chapter 5*

# IT'S A LIVIN' THING

What is it, exactly, that we are grouping when we construct trees of life or make kingdom classifications, or draw cladograms? The short answer is that we are classifying organisms, a term I've used without providing a proper definition. So let's do that here. An organism is a living thing, an entity that functions as a physiological unit, the component parts of which operate with a high degree of cooperation and a low degree of conflict to help ensure well-being and sustain the life of the overall entity and to reproduce itself so that its kind can continue.

The mission of an organism is simple: It is to acquire nutrients and energy so that growth can occur and life can be sustained to at least the point of reproduction. Organisms are born, mature, reproduce, and at some point cease to exist as a physiological unit. Only those that live long enough to reproduce have a chance of contributing to the genetic future of their kind. Viability (the ability to grow and persist) and fecundity (the ability to reproduce) are thus key characteristics of organisms. And at the heart of viability and reproduction is metabolism, the chemical process by which cells make and use energy. Indeed, Richard Dawkins describes the difference between living and nonliving things in terms of energy. Living things—organisms—make and use energy in the process of living, while nonliving things are acted on by energy—dead bodies decompose, water evaporates, rocks and minerals disintegrate.

LUCA, the common ancestor of all organisms, passed on genes that build the fundamental structure of an individual organism's constituent cell or cells. In unicellular organisms, all of the tools necessary for viability and reproduction are contained within the single cell. The job of staying alive is considerably more complex in multicellular organisms, where many different types of cells form distinct tissues, organs, and systems that must all work together as a unit. Understanding how complex multicellular organisms evolved from unicellular life-forms is a key part of our story. But so is behavior, which plays a role in evolution.

PART 2

# Survival and Behavior

*Chapter 6*

# THE BEHAVIOR OF ORGANISMS

Evolution is a process by which features of organisms within a group change over generations by means of natural selection. Underlying this process is the fit between the individual organism and its environment. Those individuals whose traits are useful in the current environment are more likely to survive long enough to reproduce and to pass their genes on to their offspring. While an organism's survival depends on many of its characteristics, as the pioneering behavioral scientist Theodore Schneirla pointed out, behavior is "a decisive factor in natural selection."

Another eminent behavioral scientist, B. F. Skinner, defined behavior as "that part of the functioning of an organism which is engaged in acting upon or having commerce with the outside world." It is common to view behavior as a product of nervous systems and the muscle contractions they control, and thus consider it as something only animals, which possess both nervous systems and muscles, do. Although Skinner's research was carried out on animals, he had the foresight to define behavior in terms of organisms in general. As we will see in the next several chapters, behavior is a feature of all organisms, not just those with nervous systems and muscles. But let's start with animal behavior.

The simplest form of behavior in animals is a reflex, an innate stimulus-response reaction that occurs by way of nerves that directly connect sensory input systems with muscles. Reflexes are automatically elicited by certain stimuli and are not dependent on volitional control. As a result, they are very useful in assessing such factors as neurological function in babies. Inborn reflex patterns are essential not only for early survival but throughout life. For example, they control balance and gait, and if you step on a sharp object, your foot reflexively withdraws. A sudden loud noise activates several reflexes, collectively called the startle response, causing your eyes to blink and neck to stiffen while your whole body flinches. When you eat, reflexes underlie chewing, swallowing, digestion, and waste excretion. As blood pressure rises, the beating of the heart reflexively slows to keep pressure in a safe range.

Reflexes typically involve the response of a specific muscle group (mouth, leg, foot, eye) or a set of responses within a body system (balance, digestion, and startle). But animals also need to perform movements that involve coordination of the entire body, and some of these are, like reflexes, inborn. Pioneering ethologists such as Konrad Lorenz and Niko Tinbergen called these more complex hardwired responses fixed action patterns, but I prefer the term *fixed reaction patterns,* which better describes their automaticity.

Like reflexes, fixed reaction patterns are elicited automatically by specific stimuli. But in contrast to reflexes, they unfold over time as complex sequences of behavior. Tinbergen's most famous example is egg retrieval by a mother goose. The sight of an egg out of the nest elicits neck extension and rolling of the egg back to the nest. Another example in birds is mouth gaping in chicks triggered by beak movements of the mother just above the nest. And in male stickleback fish the sight of a red belly (signifying an intruding male) elicits fighting while the swollen belly of a female (signifying a receptive partner) initiates mating by the male. In mammals, the sight or odor of food elicits approach, while sights, sounds, and odors signifying a predator elicit freezing, fleeing, or fighting, depending on the proximity of the predator.

Fixed reactions, not surprisingly, underlie universal survival activities, such as feeding, drinking, defense, and reproduction, and in some species, behaviors associated with parenting and social interactions as well. But the specific expression of such behaviors in a given species is unique—all animals defend, eat, and mate, but how they do so is specific to the kind of body they have. For this reason, fixed reaction patterns are considered species-typical behaviors.

In the 1950s, there were heated debates about whether these fixed behaviors are truly innate. Today it is accepted that, while there is a strong species-related (innate) component to such responses, they can also be influenced by environmental factors. For example, some complex behavioral sequences can be affected by past experience or guided and modified by current experience. Egg rolling by a goose depends in part on sensory feedback from the activity, as the response loses precision if such feedback is eliminated. Certain mating rituals are expressed only during the mating season, when environmental conditions trigger the release of reproductive hormones that lower the threshold for reproductive behaviors. Similarly, food seeking and consumption are influenced by chemical signals in the body and brain that indicate that energy supplies are low.

Although humans rely on learning and culture more than they do innate tendencies, we do still have some of these predispositions, such as our propensity to respond in certain ways to heights, snakes, the sight of an aggressive person nearby, the sound of a crying baby, and sexual stimuli. Not everyone responds to the same degree to these stimuli, but most people have at least a tendency to do so.

While reflexes and fixed reaction patterns are wired by evolution to certain classes of stimuli, both kinds of behavior can also arise, through learning, in response to novel stimuli. For example, if you haven't eaten for a while and see an advertisement for a tasty food—say, a juicy hamburger—you will likely begin to salivate. You are not salivating as a reaction to the food itself, but to a stimulus that represents the food. This is a real-life example of Ivan Pavlov's famous studies in dogs where a bell sounded before the arrival of food to elicit an anticipatory salivary reflex response (figure 6.1). The reflex response itself is innate and didn't change (wasn't learned), but the stimulus that evoked it did change (was learned to be a signifier of food). Similarly, freezing in place, a species-typical fixed reaction pattern usually elicited by the odor or sight of a predator, can come to be elicited by a novel and meaningless stimulus that is predictive of the predator. This is studied in the laboratory by Pavlovian conditioning as well, by pairing a sound, sight, or odor with an electric shock. The shock mimics the sensation of being wounded by a predator, and a stimulus that predicts the tissue irritation caused by the shock comes to evoke the defensive behavior typically elicited by a predator, as well as physiological changes that support the behavior, such as increased heart rate or faster breathing. As an actual example in humans, consider what happens if a neighbor's dog bites you as you pass his mailbox. If subsequently you are lost in thought one day while walking home and suddenly notice the mailbox, you may well momentarily freeze and notice your heart beating faster. Pavlovian conditioning is universal throughout the animal kingdom, and even exists in unicellular organisms.

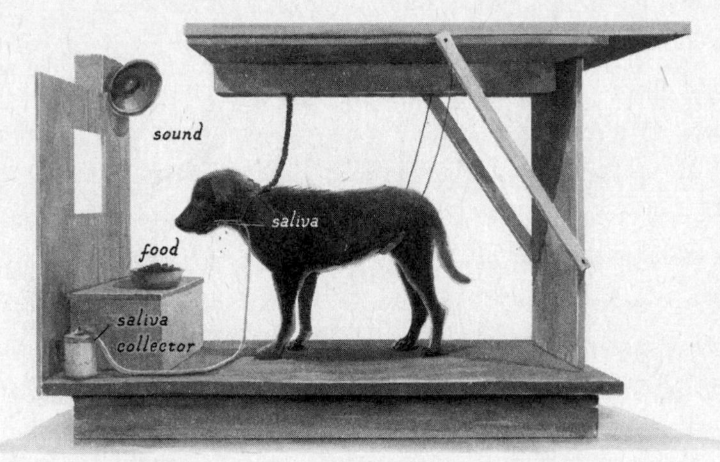


**Figure 6.1:** ***Conditioned Reflexes—Pavlov's Dogs***

The examples so far emphasize how novel stimuli come to elicit behaviors that are largely inborn. But some animals can also acquire novel behaviors through learning. If an arbitrary response, such as pressing a bar or pushing a button, is followed by a beneficial outcome (a reinforcer), it will increase the likelihood of similar behavior in the future. This effect was first described by Edward Thorndike in the late nineteenth century when he found that food-deprived cats held in a chamber became very animated when presented with food placed outside the chamber. Their random movements eventually led to their opening the door, allowing them access to the food (figure 6.2). Over the course of multiple repetitions of this, the cats learned which movement to perform to open the door. This is called instrumental learning, since the behavior is instrumental in producing the reinforcer.

**Figure 6.2: *Instrumental Behavior—Thorndike's Cats***

In the aversive domain, laboratory studies of instrumental conditioning typically involve avoidance learning, in which an animal learns to use cue stimuli (such as tones) associated with danger (typically electric shock) to guide its actions. Freezing occurs first, but once the animal learns that when the cue is present it can perform the behaviors to avoid harm, its success in doing so reinforces the avoidance response.

Sometimes the instrumental behaviors learned by reinforcing outcomes are repeated so often that they can become habits that are performed even if the response no longer produces the outcome. While instrumental habits can be useful, they also contribute to a variety of human adjustment problems—for example, continued use of addictive drugs long after their desirable consequences have ceased. Current research suggests that the outcome-dependent version of instrumental learning is limited to mammals and birds, but that habits can be learned by all vertebrates and even some invertebrates.

Stimuli connected to survival activities (defense, eating, drinking, reproduction) are the reinforcers (outcomes) that underlie Pavlovian and instrumental conditioning. This explains why in laboratory studies of instrumental conditioning, the delivery of food, drink, and sex-related stimuli, or the removal of harmful stimuli, are common reinforcers. Survival and learning are closely connected.

Some species exhibit even more complex forms of behavior that depend on cognitive capacities. Here, cognition will be used to refer to the ability to form representations and use them to guide behavior—as when you refer to a mental spatial map when planning a route and using the plan to drive to your destination. As we will see, these basic cognitive capacities, like outcome-dependent instrumental responses, are present in mammals and birds, but have not been demonstrated in other species. In mammals, cognitive capacities involving internal deliberation (mulling over what to do) are especially well developed in primates, and especially in humans. These more sophisticated cognitive capacities provide additional novel and flexible ways of meeting the challenges of survival that go beyond those provided by reflexes, fixed action patterns, Pavlovian responses, and instrumental learning. For example, humans can simultaneously hold in mind a number of alternative actions and deliberate and choose on the basis of expectations about which might lead to the most desirable outcome. The more sophisticated the cognitive capacities of a species, the greater the ability of its members to transcend survival—to act for the sake of living in a particular way, rather than merely for the sake of remaining alive. This increased capacity for cognitive control over behavior may be why fixed action patterns play such a small role in humans. Cognitive prowess also makes it possible for humans to act in ways that are potentially harmful to survival (intentionally choosing to engage in risky behaviors, such as high-speed driving, mountain climbing, skydiving, or taking addictive drugs).

*Chapter 7*

# BEYOND ANIMAL BEHAVIOR

In the nineteenth century, biologists such as Darwin and Jacques Loeb described the behavioral responses of plants to external light or chemicals. The well-known ability of a sunflower to bend to follow the passage of the sun is one such example. Primitive, stimulus-induced behaviors of non-mobile organisms such as plants are called tropisms. Plant "behavior" is a thriving area of research.

For example, in *What a Plant Knows,* the biologist Daniel Chamovitz describes sophisticated information-processing capacities that plants use to control their movements in response to stimulation. Plants not only "follow the sun" by bending their stems, they also align their leaves in such a way as to maximize exposure to light and thereby promote growth. Some plants actually anticipate sunrise from "memory," and even when deprived of solar signals retain this information for several days. In *Brilliant Green,* Stefano Mancuso and Alessandra Viola argue that plants possess not only the senses of sight, touch, smell, and hearing, but more than a dozen other sensory capacities that humans lack (including the ability to detect minerals, moisture, magnetic signals, and gravitational pull). For example, the roots of plants sense the mineral and water content of the soil and alter their direction of growth accordingly. Some plants also capture prey by sensing their presence—the most famous example is the Venus flytrap.

Some are reluctant to label plant movements as behaviors, since they lack nerves and muscles. But just as they are able to breathe without lungs and digest nutrients without a stomach, plants have the ability to move (behave). We should not dismiss the existence of behavioral capacities in an organism simply because it lacks the physiological mechanism that is responsible for the behavior in animals.

Plants clearly sense the environment, learn, store information, and use that information to guide movements; they behave. One might say that there is certain "intelligence" to their behavior. This is true as long as intelligence is defined in terms of the ability to solve problems through behavioral interactions with the environment, rather than with respect to mental capacity.

We need not, and cannot, stop with animals and plants when considering behavior. Early behavioral researchers around the turn of the twentieth century, such as Conwy Lloyd Morgan and Herbert Spencer Jennings, were very interested in single-cell protists, specifically the protozoan known as paramecia (figure 7.1). These were shown to use primitive approach and withdrawal responses called taxic behavior to engage with useful and harmful stimuli in their daily lives. Morgan, for example, noted: "The primary end and object of the receptions of the influences (stimuli) of the external world or environment . . . is to set agoing certain activities. Now in the unicellular organisms, where both the reception and the response are effected by one and the same cell, the activities are for the most part simple, though even among these protozoa there are some which show *no* little complexity of response" (italics added). The renowned early-twentieth-century British philosopher Bertrand Russell similarly pointed out: "From the protozoa to man there is nowhere a very wide gap either in structure or in behaviour." Later Lorenz concurred, suggesting that evolutionary continuity exists between protozoa and humans. Similar claims might also be made about bacteria. These ancient unicellular organisms, as discussed in the next chapter, sense and respond to their environment, and may even learn and remember.

Earlier I defined cognition as the ability to form internal representations and use them in guiding behavior. There is a movement afoot to extend the conception of cognition beyond its role in using internal representations to guide behavior so that plants and microorganisms can be

**Figure 7.1:** ***Protozoan Taxic Behavior***

said to be cognitive creatures. Some do this by equating cognition with information processing. Since all behavior involves information processing, all behavior would, under this theory, involve cognition. Others take a different tack, defining cognition as the adaptive regulation of states and interactions by an agent with respect to the consequences for the agent's own viability. With both such moves, all organisms, including protists and bacteria, are cognitive creatures.

These approaches are said to correct the anthropocentric view of cognition. But it is not the case that contemporary cognitive scientists claim that cognition is a human specialization—the study of animal cognition is a vibrant area of research. If the extended view just described catches on, it will only necessitate coming up with a new name to account for processes that underlie the use of internal representations to guide behavior. Cognition serves that role just fine. What is instead perhaps needed is a new term to characterize complex behaviors that are not under the control of internal representations.

As we proceed, I will therefore use the following behavioral categories: taxic responses, tropisms, reflexes, fixed actions, habits, outcome-dependent instrumental actions, and cognition-dependent responses. We can then call upon these as we track how features of behavior emerged as single-cell organisms were transformed into the great variety of multicellular ones that followed.

*Chapter 8*

# THE EARLIEST SURVIVORS

LUCA, the common ancestor of all life on Earth, passed on its survival-related traits in the form of genes to its descendants. Some—perhaps many—of these offspring did not persist. But one group did. By 3.5 billion years ago, bacteria had emerged, and continue to this day as the most populous kind of organism on Earth.

Whether living in a backyard garden, deep in a hot sea or the Arctic, in a muddy pond or desert, in the men's room at Grand Central Station, or serving a toxic or probiotic function in your colon, bacterial cells have to accomplish some of the same key tasks that humans do to stay alive and well: avoid danger, detect and incorporate nutrients and energy sources, manage fluids and electrolytes. And, for their kind to survive, they must reproduce. Like us, they meet many of their survival needs, in part, by behavioral engagement with their environment.

Because the entire essence of a bacterium is contained within its single cell, it does not have the additional cells required to have separate sensory organs or muscles, or a nervous system to coordinate their sensing and movement activities. The single cell is singular—on its own.

Many bacteria are motile, and use their constant random motion in daily survival. Their movements are achieved by molecular motors that control fibrous appendages called flagella. They use two kinds of movements to get around, "running" and "tumbling" (figure 8.1). When all the flagella rotate in the same direction, running produces directional motion; when the rotation is reversed, the direction is changed by tumbling.

Motility has both costs and benefits. It requires flagella, and flagella

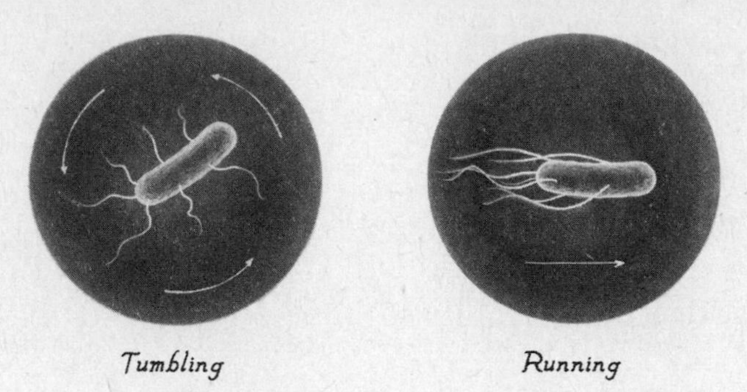


**Figure 8.1:** ***Tumbling and Running Behavior in Bacteria***

movement uses a tremendous amount of the bacterium's daily energy budget. On the positive side, motility adds leverage in food acquisition and in avoiding harm. The primitive movements of bacteria, as noted, are called taxic behaviors, orientation responses toward useful and away from harmful substances. Useful and harmful substances are called attractants and repellents, respectively. Taxic behaviors only occur in mobile organisms and are thus different from the tropisms of the stem, leaves, or roots of stationary plants; with taxic responses, the entire organism changes its spatial location.

Taxic behaviors are initiated when receptors detect attractants or repellents. Some receptors are sensitive to the concentration of chemicals in the environment (chemoreceptors) while others are sensitive to light (photoreceptors). Behaviors guided by chemicals are called chemotaxic responses and those by light phototaxic responses (figure 8.2).

Tumbling is dependent on a bacterium's momentary situation. When an attractant is detected, less tumbling occurs, resulting in running, which moves the bacterium closer to the substance. If a repellent is present, tumbling increases, changing the direction of motion, with the result being withdrawal. Whether tumbling or running occurs depends on the molecular output signals of the bacterium's receptors; the strength of these responses is dependent upon the concentration of the stimulus (chemical or light).

Bacterial cells also have to maintain the right volume of fluids within

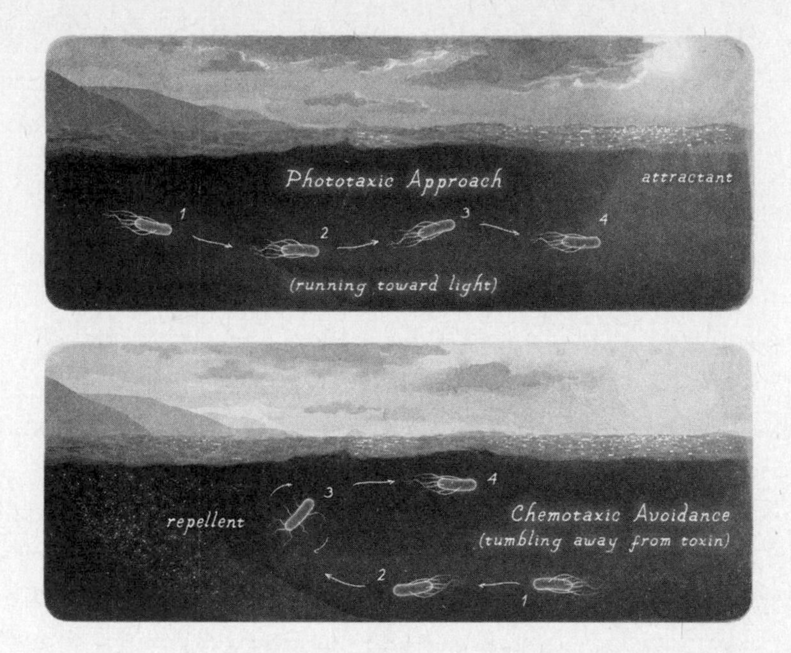


**Figure 8.2:** ***Bacterial Taxic Behavior***

the cell in order to sustain health; if there is too much water, the cell will explode, and if too little, it will implode. This process involves a complicated set of interactions between electrolytes (salts like sodium and potassium) and water. If salts are in greater concentration outside of the cell than inside, water is moved out, thus balancing the fluids and salts and preventing the explosion of the cell structure; if salts are too concentrated within the cell, water is moved in to prevent collapse.

In animals, similar factors are at play in balancing the fluids of the various cells in the organism. When our cells have too high a concentration of salt, we seek out and drink liquids to draw salt out of cells and restore balance. Fluid loss from vomiting, diarrhea, or vigorous exercise depletes electrolytes, which tips the balance in cells, and electrolytes and fluids have to be replaced, for example by consuming electrolyte-rich drinks (such as Gatorade), or in severe cases through intravenous infusions.

Many animals have a physiological process that maintains a fixed inner temperature in the face of fluctuations in external temperature, as disparity between them can be disruptive to the physiological processes and chemical reactions that are necessary for cells to survive. Animals also use behavior in thermoregulation—we strip down or find shade when too hot and add layers of clothing when too cold; some mammals shed coats seasonally; birds migrate. But rather than controlling inner temperature around a set point, bacteria maintain inner temperature by reconfiguring certain biochemical processes so as to adjust their physiology to match the external temperature. This capacity may have been a key to the ability of early unicellular life-forms to survive in a variety of climates. But additionally, bacteria sense external temperature and use behavior in accommodating to the outside world.

Like all organisms, bacteria have to reproduce in order to persist over time. For us, and many other animals, reproduction typically involves another organism. But for bacteria, it is simply a matter of cell division. Sex arose as a modification of cell division in single-cell protists and would not exist without the asexual solution to reproduction that bacteria have used for billions of years.

Although reproduction itself isn't a social activity for bacteria, they are, in some sense, social organisms, often collecting together on surfaces and exuding chemicals that cause the individual cells to literally stick together. Such collections are called biofilms. The plaque on your teeth, the goo on your shower wall or around your electric toothbrush, and many scummy-looking things in nature are instances of biofilm. Recent studies have shown that these cells are not just passively attached but actually communicate by generating electrical signals, which they use to coordinate feeding and reproduction and to attract new members to the group. Many behaviors that pass for psychological or social interactions in more complex organisms, like attraction to and avoidance of places and things based on odors, are actually accounted for by similar simple factors.

Bacteria can also acquire information about their world and use it in the future to guide behavior. For example, evidence exists that they form internal molecular representations of environmental conditions (temperature, oxygen levels) and use them to predict environmental conditions later so that they can respond appropriately. In other words, as noted, they may learn and remember. While the evidence for learning in bacteria is largely based on theoretical models, solid evidence of learning and memory is present in unicellular protists. Learning and memory do not require a nervous system.

We often take an anthropocentric view of memory, equating it with our ability to consciously recollect the past. But memory is first and foremost a cellular function that facilitates survival by enabling the past to inform present or future cellular function, whether in a single-cell or multicellular organism. The same is true of much of the rest of our psychological life and its manifestations in our conscious minds. Consciousness, though useful to humans in ways that will be discussed later, is often a passive observer of behavior rather than an active controller of it, especially with regard to survival mechanisms that originated billions of years ago. It is this ability of our brains to be consciously aware of their own activities that allows us, when we are in a survival state, to experience fear, pleasure, and other emotions, and even to exercise free will in choosing what to do next.

*Chapter 9*

# SURVIVAL STRATEGIES AND TACTICS

Ancient unicellular organisms solved the initial problems of survival by establishing mechanisms necessary to sustain life. With the emergence of more complex organisms, the survival wheel did not have to be reinvented. It just had to be modified. Its specific implementations in particular species, though, were determined by the conditions under which each species arose.

For example, different animals have visibly different kinds of bodies. These reflect the unique ways that their ancestral members of the species adapted to their environment. As a result, different animals express different outward behaviors in survival situations: some swim, others fly, and still others move over land by slithering or by locomoting on four, or even two, legs. The specific implementation required varies considerably depending on the body plan that the animal possesses.

Evolutionary biologists distinguish ultimate from proximal explanations. The answer to the question of "why" a behavior exists is the ultimate, or evolutionary, explanation, while the answer to the question of "how" the behavior is expressed in an existing organism in a given situation is the proximal explanation. The proximal reason that an animal runs from a predator is because its brain processes sensory signals about the predator in such a way as to initiate protective actions, such as escape. The ultimate reason the animal does this is because it belongs to a lineage of organisms that were adept at escaping from harm, and lived to pass genes for this trait on to their offspring.

The biologist Karl Niklas argues that in the process of creating new species, natural selection "acts on functional traits rather than on the mechanisms that generate them." In other words, conserved genes translate into the general functions that keep an organism alive, and not their specific implementations. Unique implementations in a species require novel mechanisms.

One way to think about all this is in terms of strategy and tactics. Survival strategies reflect the basic requirements of cellular life and are universal. Because different organisms have different physical structures that reflect ways that the body plan of their species has been tailored to its environment through natural selection, they possess different behavioral tools or tactics to implement the survival strategy.

The most obvious method by which survival activities might have been maintained throughout the history of life is through conserved genes. Modern genetics has provided compelling evidence that there exists a ubiquitous set of genes, bundled together as a so-called universal genetic tool kit, that is shared between unicellular and multicellular organisms, some of which were present in LUCA. This fact alone is astounding: all living creatures have genes that date back to the beginning of cellular life. At this point, however, we don't know whether genes underlying the ability of early organisms to maintain energy supplies, balance fluids, and defend against harm have been conserved. But the scientific tools exist to pose and answer the question.

Making a leap from the simple behaviors of unicellular organisms to the more complex ones of animals might seem far-fetched. But it becomes more plausible when we consider that most behavior involves movement toward or away from environmental stimuli. Approach and withdrawal, as pointed out by Theodore Schneirla, are the most fundamental, universal classes of behavior.

But why are they universal? Given two stationary objects, movement of one will cause it to be either closer to or farther away from the other; there are no other possibilities. Approach and withdrawal may be universal simply because they are subject to the laws of physics, not because they reflect psychological motivations.

But it is also the case that organisms that were more effective in moving toward environmental elements that support survival (attractants) and away from elements detrimental to survival (repellents) would have a selective advantage, and likely pass the advantageous genes on to their offspring. But what they pass on is not so much the physical tendency to approach and avoid, but specific biological implementations related to survival needs.

For example, the evolution of nervous systems in animals provided novel options for approach and withdrawal beyond simple taxic responses of unicellular organisms. In particular, neural control allowed approach and withdrawal to be implemented in more precise ways in the service of specific survival requirements. As animals and their nervous systems became more complex and sophisticated, so, too, did their behavioral repertoires. When organisms satisfy specific survival needs related to energy and nutrition, fluid balance, and defense, or when they engage in sexual reproduction, they are, in the most basic sense, typically using neural circuits to call upon specific approach and withdrawal behaviors that are matched to their body plan by their species' genes.

We often attribute mental states, especially emotions, to behaviors associated with survival. We say we withdraw from danger because we feel afraid of what will happen. If withdrawal fails, we say we feel dread or terror. Similarly, our approach to food or sex is often said to be accompanied by desire or hope. When approach activities succeed, we experience satisfaction or pleasure, and disappointment or frustration when they fail. While there is no denying that we experience such emotional states, we need to be cautious about calling upon them to explain behavior. As I will argue later, research suggests that approach and withdrawal and other survival behaviors in humans are mediated by different brain circuits than those that result in feelings of fear, pleasure, disappointment, and so on.

We know not about the feelings of other animals. But the issue here is not what other animals experience. It is instead about where our survival activities come from. They are products of the long history of evolutionary adaptations that allowed LUCA's descendants to survive and ultimately evolve nervous systems to accomplish survival activities.

The key point of this chapter is that we and other complex organisms have to do some of the same general things to survive that simple organisms do to stay alive. But the implications of this conclusion should not be overinterpreted. It is simply a statement about the universal, minimal requirements of life. It is not meant to trivialize the sophistication and complexity of the behaviors of humans and other animals, but rather to emphasize the sophistication of cellular survival activities in simple cells, and the deep connection of these ancient survival activities to our lives today.

*Chapter 10*

# RETHINKING BEHAVIOR

Behavior, especially human behavior, is often treated by scientists and laypeople alike as a special kind of response, one connected not just with the body but also with the mind. This common view of the relation of behavior to the mind is called folk psychology, and is reflected in everyday vernacular language. But as the psychological historian Kurt Danzinger pointed out, everyday language and the mental conceptions it gives rise to emerged long ago, somewhat arbitrarily and out of convenience, and is poorly suited as a framework for evaluating scientific concepts. Danziger contends that it would in effect have been a miracle if ancient people just happened to have hit upon the perfect way to talk about what goes on in our brains and minds.

Our everyday vernacular language arose and persists because it enables discourse about the inner lives of people as they interact with one another. It has been tremendously advantageous for our species to have this ability. Without it, human culture as we know it would not exist. But the effectiveness of the vernacular in daily life does not necessarily carry over to the realm of scientific discourse. The key issue is when, if ever, the vernacular provides an accurate vocabulary for pursuing scientific discoveries about mind, behavior, and the brain.

While we humans easily convince ourselves of the connection of mental states to behavior, it is, in my view, a convenient illusion, if not a delusion, at least some of the time. As noted above, different brain circuits seem to underlie the conscious feeling of fear and the behaviors that occur when one is in danger. This doesn't mean that fear never affects our behavior, but rather that it is *not* a foregone conclusion that what appears to be fear behaviorally is, in fact, due to an underlying feeling of fear that one is experiencing. If the connection between behavior and mental states is somewhat flimsy in humans, it is on even weaker footing in other organisms, since we have no way of knowing with any certainty what the mental life of other creatures might be like.

Scientists sometimes use everyday words as convenient terms for scientific entities. In physics, particles are named WIMP, GOD, and quark, while biology labels certain genes and molecules dunce and hedgehog. No one takes such names too literally. But when neuroscientists and psychologists describe a brain circuit that controls behavioral responses to danger a "fear" circuit, some of them (and most laypeople) believe that this circuit is the source of the experience of fear. In other words, when scientists use mental-state words to describe circuits that control behavior, they effectively ascribe the mental-state function named by the word to the circuit itself, even if their intention was to simply use "fear" as a convenient label for the circuit. One consequence is that over time the more subtle distinctions are lost, even among scientists, and efforts to identify the circuit that actually gives rise to the subjective experience of fear ceases, since it is assumed that the behavioral control circuit is responsible for the feeling.

In the early twentieth century, a movement arose in psychology in which all talk of mind and consciousness was eliminated. This behaviorist revolt was meant as an antidote to the rather casual way that scientists had been attributing humanlike thoughts and emotions to animals when explaining why certain behaviors were performed. While the behaviorists developed excellent experimental methods (like Pavlovian and instrumental conditioning), which are still used, their radical ideas about a "mindless" psychology did not have staying power.

Still, some philosophers and scientists continue to argue that science will eventually replace quaint folk-psychological notions. I think they are right about most folk-psychological notions about behavior, but wrong about the folk psychology mental states like thoughts and emotions. As Harold Kelley and Garth Fletcher have independently argued, everyday folk psychology has a purpose and will always be a part of the psychological effort to understand our inner subjective experiences. Folk psychology often serves as the starting point for scientific inquiry into mind and behavior. And because folk psychology is entwined with our vernacular language regarding the mind, our mental-state words provide labels for meaningful categories of experience. We may modify or eliminate certain category labels, but that is different from eliminating the underlying psychological experience.

Because subjective experiences are real and play important roles in human life, mental-state terms are needed to describe them scientifically. But to do their job conceptually, their use must be restricted to mental states. We have to be more diligent about separating those aspects of behavioral control that involve such conscious experiences from those that do not.

Behavior did not arise to serve the subjective mind. It came about and persists to enhance fitness—to keep organisms alive and well so that reproduction can occur. This perspective puts the behavior of humans and bacteria, and all organisms in between, on a level playing field, one in which consciousness, in the sense that humans mean by the term in everyday life, has a peripheral role in most of the history of life. If we commit to the very reasonable assumption that over the long course of evolution behavior has mostly been generated by nonconscious systems, and that behavior, even in humans, should be assumed to be nonconsciously controlled unless proven otherwise, the science of behavior would advance much more smoothly. And, by the same token, so would the science of consciousness. This is the perspective I will take as I guide you on our ascent of the tree of life in the remainder of this book.

## Part 3

# Microbial Life

*Chapter 11*

# IN THE BEGINNING*

The universe began some 13.7 billion years ago. Its expansion led to the formation of galaxies and stars about 10 billion years ago, and of the sun, Earth, and the rest of our solar system 4.6 billion years ago. By roughly 3.8 billion years ago, life had begun on Earth (figure 11.1). How did this happen?

**Figure 11.1: *The Big Bang***

Origin-of-life theories, of which there have been many, propose explanations of how the early physical chemistry of the Earth was transformed into the chemistry of life—how living organisms sprang from nonliving matter. It all starts with the "big bang," from which Earth inherited its initial chemical makeup. The known chemical elements on Earth—that is, the elements in the periodic table—are essentially traces of stardust

* Thanks to Tyler Volk for suggestions on this and other chapters in part 3. His 2017 book, *Quarks to Culture*, provides an excellent summary of the physical and biological factors that account for life and its many manifestations, including culture.

from the "big bang" and subsequent supernova explosions. Literally everything natural on Earth, whether living or nonliving, is composed of chemicals that arrived cosmically, including the air you breathe, the land you walk on, the water you drink and swim in, the plants and animals you eat or keep as pets, and the body, brain, and mind you use to do all these things.

When initially formed, the Earth was a hot, molten mass. But by 4.2 billion years ago the surface had cooled down enough to form a solid crust surrounded by a primitive atmosphere of mostly carbon dioxide, water vapor, and nitrogen. The oceans were formed by steam arising from volcanoes and water from incoming meteorites. There was no life at this point, but some of the necessary requirements for it, including water, a supply of carbon, and heat to facilitate chemical reactions, were available.

Water provided a medium in which chemical compounds could dissolve and be reconfigured to make other compounds. For example, if you pour table salt into a glass of water, it easily dissolves—oxygen and hydrogen molecules from water attract sodium and chloride, respectively, pulling the salt molecule apart.

Carbon is a small atom that easily interacts with (chemically bonds with) other small atoms to form compounds. These carbon-based compounds, which are the foundation of living matter, are relatively stable and do not readily dissolve in water. (We don't come apart when we swim.) But if the surrounding temperature increases, their breakdown is facilitated. Sources of heat were readily available on the young planet in the form of sunlight, volcanoes, underground magma in the Earth's mantle, and bolts of lightning striking water. The constituent atoms resulting from the heat-facilitated dissolution of carbon compounds thus became free to be recombined.

Darwin used such information about the early world of carbon-based molecules to propose that life began in some primordial pond when lightning caused nonliving carbon compounds to be dismantled and reassembled into other compounds that could undergo more complex configurations typical of life. The biologist J. B. S. Haldane expanded and popularized this idea, which came to be known as the "primordial soup theory" of life's origins.

But the early physical chemistry of the Earth, even with the addition of carbon-based chemistry, would not likely have been sufficient to make life possible in Darwin's scenario. The Israeli chemist Addy Pross argues that life required a rewriting of the chemical rules. He calls the chemistry of the prebiotic world (the world before biology, or life) regular chemistry, and that which formed the basis of life, replicative chemistry.

Through regular chemistry, atoms can be combined to make complex carbon-based molecules. With just 6 of the 118 elements in the periodic table (carbon, hydrogen, nitrogen, oxygen, phosphorus, and sulfur), each of which was available in the prebiotic world, carbon-based molecules that play crucial roles in cells, the basic unit of life, could be made. These included proteins, lipids, and carbohydrates (table 11.1). Proteins are the foundation of cell function. They act as receptors that detect useful chemicals or transport chemicals in and out of the cell, or as enzymes that regulate chemical reactions, such as those underlying metabolism (the energy-making capacity of cells). Proteins also contribute to forming an internal scaffold that gives the cell shape. While carbohydrates also contribute to cell structure, they are most important for their role in energy metabolism and energy storage. Lipids form the casing or membrane that surrounds the cell and separates it from the outer world, and also store energy and provide insulation (think of fat).

TABLE 11.1: Elements Contributing to Biological Molecules

| BIOLOGICAL MOLECULE | ELEMENTS | EXAMPLE |
|---|---|---|
| Carbohydrates | C, H, O | glucose: $C_6H_{12}O_6$ |
| Lipids | C, H, O | cholesterol: $C_{27}H_{45}OH$ |
| Nucleic Acids | C, H, O, N, P | DNA (base A): $C_{10}H_{12}O_6N_5P$ |
| Proteins | C, H, O, N, S | insulin: $C_{257}H_{383}N_{65}O_{77}S_6$ |

Replicative chemistry also involves interactions between carbon-based compounds—in this case, nucleotides. These consist of the same elemental atoms as the organic compounds mentioned above. But compounds that result from nucleotides are extraordinary. They are called nucleic acids, which you know as DNA and RNA, chemicals that can self-replicate. And with self-replication, we are poised to consider how replicative chemistry made life possible.

*Chapter 12*

# LIFE ITSELF

Cells with the ability to sustain life and replicate did not arise out of the blue. There were surely false starts and dead ends in biological experimentation before a cell came along that could sustain life and reproduce itself.*

One way that scientists have tried to figure out how sustainable cells came to be is by attempting to simulate the early chemistry of a primordial pond or ocean. The most famous example is an experiment performed by Stanley Miller, working in the Harold Urey laboratory in the 1950s. Miller put chemicals that he thought might have been present in the primordial atmosphere (hydrogen, ammonia, and methane gases) in water and passed electricity (simulating lightning) through the mixture, hoping to trigger the conversion of prebiotic carbon-based compounds into biological compounds (figure 12.1). Several days later Miller found that amino acids, which are the building blocks of proteins, a key ingredient of life, had formed, demonstrating that inorganic elements, in the presence of heat, can form biological compounds. Later research suggested that the early atmosphere of the Earth probably involved different gases than those used by Miller. But the study showed that understanding the transition from prebiotic chemistry to biochemistry was potentially within reach using scientific methods.

In recent years the debate has focused on which of two key events came first. The choices are metabolism (which provides the energy necessary for a biological entity to persist during its lifetime) and replication (which

* Michael Marshall's 2016 article on the BBC website gives a cogent summary of beginning-of-life theories (for details see the bibliography for this chapter).

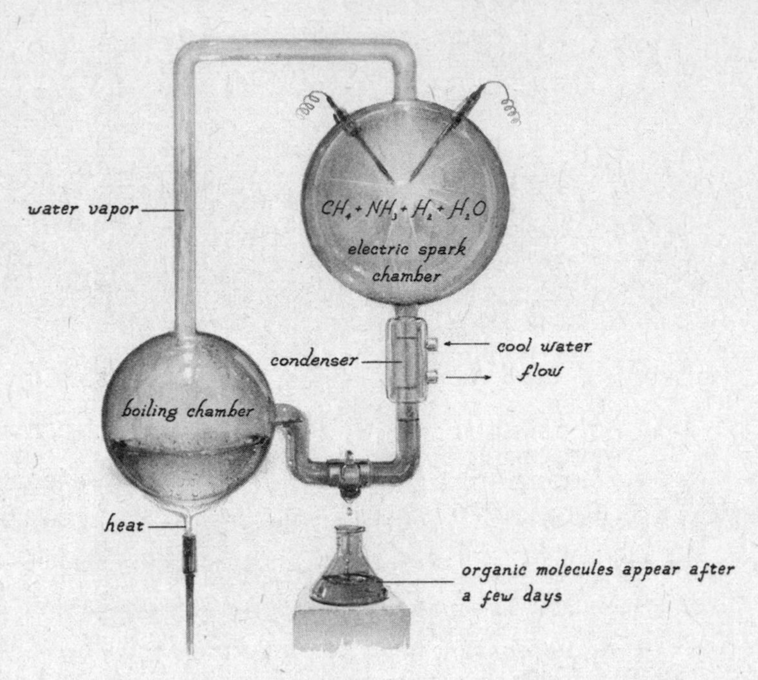


**Figure 12.1:** ***Miller and Urey's Experiment to Create Life in the Lab***

makes it possible for an entity to duplicate or reproduce itself and thus transcend individual life). The replication-first theory argues that self-replicating prebiotic molecules paved the way, and ultimately led to biological replication, with metabolism following. All a prebiotic molecule has to do in order to support replication is make new copies faster than the old ones degrade. One suggestion is that replicating carbohydrate polymers (complex sugars) were able to do so. While the chemistry of Earth at the time probably couldn't support these reactions, interstellar chemistry could. So it's possible that "space sugars" contributed to the emergence of biological replication in the form of nucleic acids.

Biological replication, as we know it today, depends on the coding of the genome by DNA. But some biologists, like Gerald Joyce, believe that replication depended on RNA before the development of DNA (figure 12.2). RNA would have been sufficient to jump-start biological evolution, but would not have been able to sustain life because it has limited

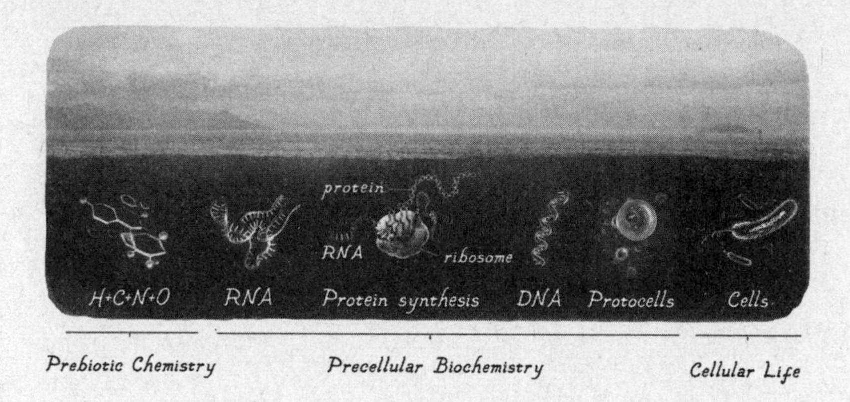


**Figure 12.2:** ***The RNA-First Theory of Life's Origins***

stability and cannot support a large genome. DNA was the solution to those constraints.

DNA is believed to have emerged through a transformation of RNA, possibly by a virus that converted an RNA gene into a DNA gene. That this is plausible has been shown in studies that demonstrated the transfer of genes between RNA-based and DNA-based viruses in acidic water, a milieu that is thought to be similar to that of the early oceans in which life arose.

Self-replicating RNA and DNA molecules, according to the replication-first theory, were initially free-floating, but were later compartmentalized, a step that would have offered the advantage of allowing the protein products of RNA and DNA to be confined within, and thus used solely by, the structure that housed them.

Initially compartmentalization might have been accomplished by a nonbiological transitional entity, a protocell, perhaps formed within pores in rocks. (We'll consider this in more detail below.) But a protocell in a rock pore, even one containing DNA, would not be capable of sustaining complex life. The evolution of true cells depended on some form of compartmentalization outside of such confined spaces. The eventual solution was a lipid casing (membrane) that sequestered RNA, DNA, and the proteins they make, allowing these entities to exist free-floating in the oceans, where they could self-replicate, diversify (that is, evolve), and give rise to all of the organisms that have ever lived.

The other theory argues that replication based on nucleic acids (RNA and/or DNA) came after biological entities could support metabolism. Günter Wächtershäuser proposed a version of this metabolism-first theory in which hot water from volcanoes flowed over mineral-rich rocks to ignite (catalyze) chemical reactions that fused simple carbon-based compounds into larger ones. While catalytic enzymes, which are proteins, did not yet exist, minerals, such as those in rocks, can and do function as prebiotic catalysts for chemical reactions. According to this theory, a key step occurred when, through a series of these prebiotic reactions, the circle was closed by the regeneration of the original compound. Through such a process, complex biological molecules (proteins, nucleotides, lipids, and carbohydrates) could be made, forming the basis of simple protocells that made energy and replicated.

Wächtershäuser's idea was criticized on the basis that the heat in volcanic flows, as described above, would have been too high to allow life to exist for more than a few seconds. A variant of the metabolism-first theory, called the alkaline hydrothermal vent theory, first proposed by Mike Russell and Bill Martin, and later expanded upon by Nick Lane, solved this problem (figure 12.3).*

**Figure 12.3:** ***The Hydrothermal Vent Theory of Life's Origins***

* I am grateful to Nick Lane for helpful comments, especially in constructing the illustration of the hydrothermal vent theory. However, I take responsibility for simplification that appears in the depiction.

It's a fascinating but complicated idea, explained in the following simplified synopsis: The starting point is the widely accepted assumption that the early oceans were cold and acidic (had a low pH) due to an abundance of positively charged chemicals. Through cracks in the ocean floor, the cold, acidic water seeped down into the Earth's mantle, where it was heated and flowed back up to the ocean through mineralized vents made of carbonate and pyrite. In its passage through the vents, the warmed liquid acquired an alkaline nature (high pH, and thus a negative charge)—similar to the way a carbonate compound in Alka-Seltzer counteracts the burning sensation from stomach acid by increasing pH. This alkaline fluid was trapped in the pores of vents and separated from the positively charged surrounding ocean water by a foam barrier consisting of iron sulfide.

Only a few elements were needed to form protocells. Hydrogen (H) entered into prebiotic chemical reactions with carbon dioxide ($CO_2$) to construct simple carbon-based molecules, such as formaldehyde ($CH_2O$) or acetate ($CH_3O_2$). But hydrogen enters such reactions only in the presence of some sort of catalyst. Again, protein-based catalytic enzymes did not yet exist, but other catalysts did. In particular, iron sulfide, which was the basis of the proposed foam barrier in protocells, is an excellent prebiotic catalyst, and is still used in metabolism by cells. Metabolic cycles, such as those proposed by Wächtershäuser, could then store energy to create more complex biological compounds, including proteins, lipids, and replicating nucleotides, in the pore-based protocell. When the protocell eventually acquired a lipid membrane, it could leave the pores and lead a metabolically self-sufficient and replicative life in the ocean.

Whether replication or metabolism came first, or somehow the two evolved in tandem, remains to be determined. Regardless, these theories have identified plausible explanations for why every cell, from LUCA forward, is surrounded by a lipid membrane, has an inner negative charge (a fact that became very important when neurons emerged in early animals some 3 billion years later), uses metabolic cycles to generate and store energy, and replicates itself by making proteins under the instruction of DNA.

An alternative view of life's beginnings was proposed recently by Leroy Cronin and Sara Imari Walker. Advocates of replication and metabolism theories debate the order in which metabolism and replication appeared, but assume that replication by nucleic acids preceded life. Cronin and Walker argue that this would have required complex events with low probabilities to have transpired. Suppose, however, that instead of building RNA and DNA from prebiotic elements, information in simple networks simply copied itself, allowing life to have started as a sudden change in how regular physical chemistry processed and used information. RNA and DNA, rather than being the starting point, would have appeared later as refinements of how such information was processed. While this is an interesting possibility, the early arrival of RNA and DNA is still the dominant view, and the view we will assume is correct as we climb and explore the branches of the tree of life.

*Chapter 13*

# SURVIVAL MACHINES

Life as we know it began roughly 3.8 billion years ago and depended on the cell known as LUCA surviving long enough to replicate. By 3.5 billion years ago, LUCA's descendants had diverged to form what we now know as the kingdom occupied by Bacteria. Archaea then soon branched off to form a second kingdom. Bacteria and Archaea have been in the business of life for a very long time and have, so to speak, seen it all, and lived to tell the tale. We can learn a lot about life from these ancient survival machines.

One of the reasons for their success, and hence longevity, is that they are able to live under many different climatic conditions. Bacteria exist everywhere on Earth—on land, in the sea, and in the air. They survive well in the moist, warm recesses of our bodies (where we have many more bacterial cells than human cells) but also in snow and ice, and in the magma-heated high-temperature waters of deep-sea vents. Archaea are especially impressive in that they can survive at temperatures upward of 200 degrees Fahrenheit, in waters of high salt concentration (ten times that of the normal amount in most areas of the ocean), and even in acid. Cells that tolerate extreme environments are called extremophiles.

The fluid-filled interior of a cell is called the cytoplasm (figure 13.1). It makes up the body of a cell and is enclosed in the cell membrane, which separates it from the outside environment. A version of the lipid membrane that evolved in early cells is still used by all cells to compartmentalize metabolism and replication processes.

The survival and well-being of any cell depends on the active exchange

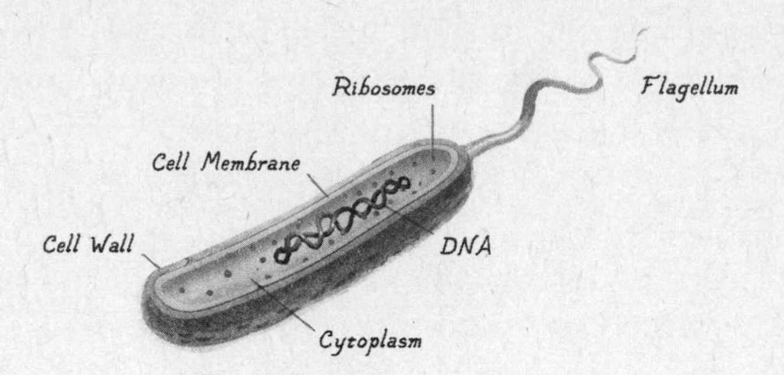


**Figure 13.1:** ***Inside a Bacterial Cell***

of molecules between the outside world and the cytoplasm, by passage through the cell membrane. The membrane is selective in what it lets inside the cell. This discretionary permeability is dependent on the molecular composition of the membrane—two layers of lipids and small spaces between segments. This configuration allows some molecules, like water and nutrients, to enter the cell through the spaces relatively easily, while other molecules need help getting in and out, and are chaperoned by so-called transporter proteins.

Once inside the cytoplasm, externally acquired ingredients contribute to the complex chemical reactions that make enzymes and other proteins used in energy generation, maintenance of fluid and ion balance, regulation of inner temperature, and control of cell movements in acquiring nutrients and defending against harm. Metabolism generates wastes that must be excreted through the cell membrane, often with the help of transporters as well.

In bacteria and archaea (and also plants), the cytoplasm is surrounded by an additional layer of protection called the cell wall. Unlike the membrane, which serves as a filter, the cell wall allows many more chemicals, all except the largest, which are often toxins, to move freely through it. Its rigid shape prevents the cell from collapsing when water leaves, and from exploding when it enters.

Another important function of the semipermeable nature of the cell membrane is the maintenance of the balance in charge between the inside and outside of the cell. Recall that since the beginning of life in the oceans, cells have been configured so that their interior is more negative than the acidic (positively charged) surrounding environment. This chemical balance between the inside and outside is essential for sustaining the inner workings of metabolism.

Tyler Volk points out that cells are self-generated dynamic entities that at any given moment are always on the cusp between persisting and perishing. They manage to survive by using their metabolism to stay ahead in this game. When metabolic wastes are expelled, the result is a loss of molecules. To compensate, cells also use metabolism to grow new molecules. If the exchange is at least equal, the cell can persist in its present form. If more molecules are generated than are lost, which adds protection against perishing, net growth results, and the cell gets bigger. But a cell can only grow so much, as larger cells require more nutrients, and the cell runs up against a basic principle of physics—as a sphere gets bigger, its interior increases to a greater degree than its surface area. For a cell, this makes it harder for the surface to keep the flow of nutrients high enough to sustain the ever-larger interior. So what's a cell to do? It divides in half and starts the process all over as it approaches its useful size limit. This achieves a balance between growth and persistence.

Indeed, bacteria and archaea replicate by simple cell division. This is asexual reproduction, since only one organism—one cell, in this case—is involved (later, in chapter 18, we'll look at differences between asexual and sexual reproduction). When the cell reaches a certain point in its life cycle, its genes duplicate (replicate), and the two resulting complete genomes segregate, one on each side of the cell. The cell then splits in half, a process called mitosis, forming two daughter cells, each of which gets identical genes. In a sense, the first bacterial cell that ever lived is immortal—it continues to divide in every existing bacterium.

The passing of genes from parent to offspring is called vertical gene transfer. Given that both daughter cells resulting from mitotic cell division have the same genes, do they go through life as identical twins? In fact, bacteria and archaea have considerable genetic individuality as a result of horizontal gene transfer, a process by which genes are acquired from other organisms (figure 13.2). For example, cells randomly deposit genes in their surroundings, and other cells can pick them up. This adds genetic diversity, even between cells that have the same parent. In addition, bacterial and archaeal cells, like all cells, can also undergo both beneficial and harmful mutations, adding further genetic diversity. And when a bacterial or archaeal cell divides, all of the genes in its possession, including mutated ones and those it inherited vertically from its mother cell, as well as those picked up from the environment, are passed on.

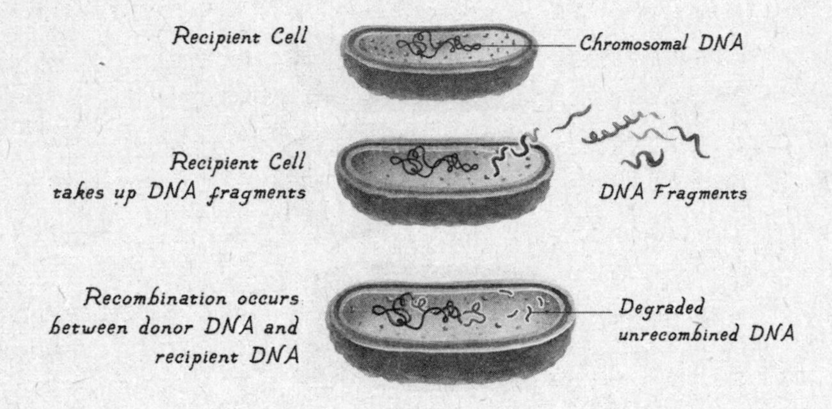


**Figure 13.2:** ***Horizontal Gene Transfer in Bacteria***

Bacteria have an uncanny way of surviving adversity. A major adversary these days is antibiotic drugs. These combat bacterial infections by damaging the cell membrane, which results in disruption of the balance between the inside and outside of the cell, and a leakage of proteins and other resources that are essential to its survival, and thus cell death. As shown in a video made by Michael Baym, and in an article in *The Atlantic Monthly* online, Baym created a very large petri dish in which he created concentric circles of antibiotic, with the dose of the antibiotic progressively stronger in rings closer to the center of the dish. Bacteria were then placed around the dish's edge, where there was no antibiotic, and videoed continuously for several months while they divided many, many times (in general, each cell divides several times per hour) and moved in relation to the antibiotic rings. Initially the bacterial group as a whole began avoiding the outermost ring. With more and more cell divisions, new generations resulted that could survive the first dose, and these moved into the second area, and on and on, and eventually completely covered the dish. This showed, in real time, how antibiotic-resistent strains of bacteria emerge.

Bacteria thus are very good at surviving. Without a nervous system, they can overcome even the cleverest challenges our nervous system can devise to thwart them, at least so far.

*Chapter 14*

# THE ARRIVAL OF ORGANELLES

The six kingdoms of life are sometimes grouped into three domains (figure 14.1). The first two are composed of Bacteria and Archaea. The third consists of all organisms that are not members of Bacteria or Archaea—in other words, all members of the Protista, Plantae, Fungi, and Animalia kingdoms. The name of this domain is Eukarya. While bacteria and archaea are unicellular organisms, eukaryotes come in both unicellular (protist) and multicellular (plant, fungi, animal) varieties.

Bacteria and archaea had the Earth to themselves from around 3.5 to

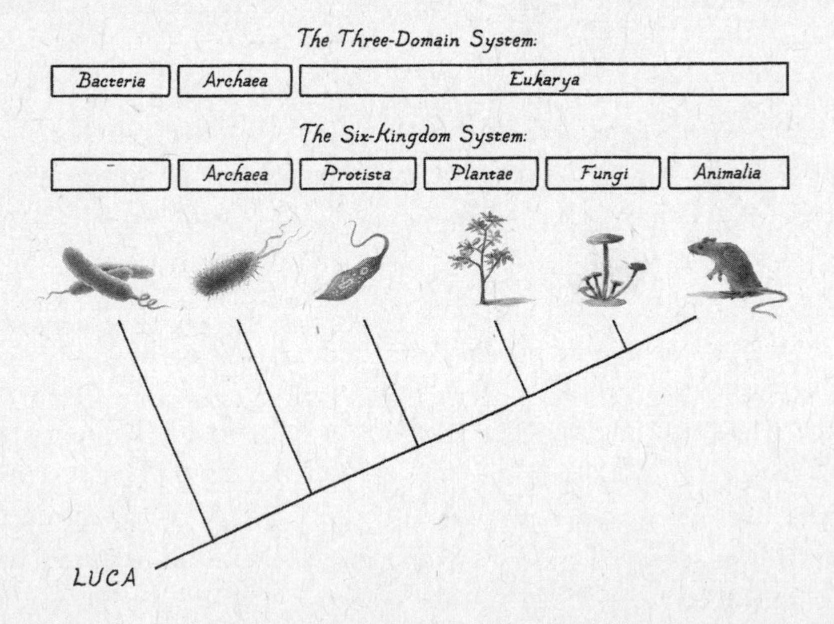


**Figure 14.1:** ***Kingdoms and Domains of Life***

2 billion years ago. Then, suddenly, they had to share the planet with eukaryotes. The first eukaryotes were, like bacteria and archaea, single-cell microbes. But they differed from their predecessors in several fundamental ways.

All cells have a cytoplasm surrounded by a cell membrane and DNA in the cytoplasm. In bacteria and archaea, DNA floats freely within the cell, but in eukaryotes, it is isolated from the rest of the cytoplasm (figure 14.2). The word *eukaryote* literally means "true kernel" or "compartment" (from the Greek: *eu* = true + *karyo* = kernel). The DNA-containing kernel is called the nucleus. Bacteria and archaea are called prokaryotes, which literally means "before the kernel."

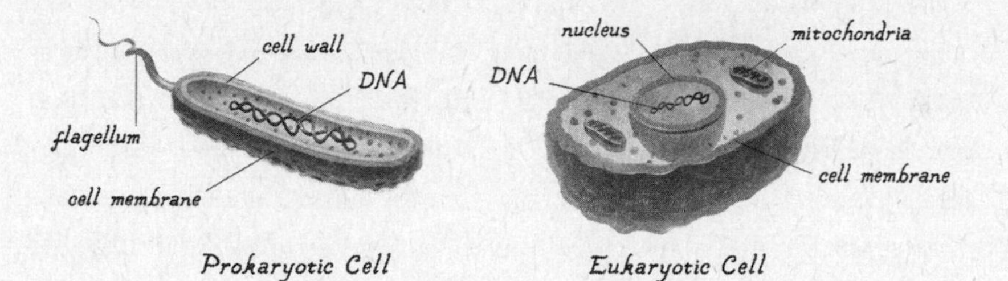


**Figure 14.2: *Prokaryotic vs. Eukaryotic Cells***

The eukaryotic nucleus is surrounded by a membrane, the nuclear membrane. Membrane-enclosed structures like this that exist in the cytoplasm of eukaryotic cells are called organelles (little organs). Another type of organelle is mitochondria. These are dedicated energy machines that are more efficient than the energy-producing capacities available to prokaryotes. Other structures in the eukaryote cytoplasm are the endoplasmic reticulum and Golgi apparatus, which are involved in the manufacture and regulation of proteins made under instructions from the DNA in the nucleus.

Unlike prokaryotes, most eukaryotic cells do not have a cell wall; they are only enclosed within the cell membrane (plant cells are an exception, as they have a wall and a membrane). Unlike the cell wall, the cell membrane is not rigid. Thus, cell structure must be maintained in most eukaryotes by another means—an elaborate system of filaments, made of proteins, that form an inner scaffold called the cytoskeleton (cell skeleton) is used. Prokaryotic cells also have a cytoskeleton, but it is less elaborate, as it does not have the burden of maintaining cell shape.

The eukaryote cytoskeleton has another important role—it constitutes a chemical transport system that facilitates communication between the different parts of the cell. This is an important feature since, as we will see later, eukaryotic cells are larger than prokaryotic cells, and some means of chemical communication was needed to enable efficient communication between their various parts. But they also needed other changes related to metabolism to support their larger cell volume, which is the role that mitochondria play.

Another key difference is that eukaryotic cells evolved in such a way as to give rise to the development of macroscopic (visible to the naked eye) multicellular organisms (plants, fungi, animals)—a feat that prokaryotes never achieved. Still, microscopic prokaryotes win out decisively in terms of the sheer number of individual organisms. It is estimated that there are currently $5x10^{30}$ bacteria and archaea alive, which is too large a number even to have a moniker like million, billion, or trillion.

The last difference we'll mention here between the domains is by no means the least important. While prokaryotic cells reproduce by simple cell division, eukaryotes invented sexual reproduction, in which two mating types, or sexes, are required, one of which fertilizes the other. We'll expand on some of these differences between prokaryotes and eukaryotes as we proceed in this section and the next.

*Chapter 15*

# THE MARRIAGE OF LUCA'S CHILDREN

It's generally accepted that the evolution of eukaryotes involved two key modifications of prokaryotes. One, obviously, was the emergence of the cell nucleus. This is believed to have occurred by "infolding" of the cell membrane in archaeal cells, a process in which part of the membrane became detached and coalesced around the gene-carrying chromosome. The cell nucleus thereby became the command center of the cell (DNA instructs protein synthesis by RNA, and all cell functions depend on proteins).

Although the presence of a nucleus is what gives eukaryotes their name, and was certainly evolutionarily important, another event was also necessary for the emergence of eukaryotes as a new life-form. This was a marriage of sorts, a form of cellular incest between LUCA's descendants.

It was in some respects a forced marriage, as it involved an archaeal cell engulfing (basically, swallowing) a bacterial cell (figure 15.1). It's not uncommon for organisms to consume one another—we eat plants and other animals, for example. But in the case in question, the bacterial cell successfully avoided being digested—just as intestinal parasites manage to live inside and off of their animal hosts without being consumed. While it is possible that the engulfed bacterium was initially parasitic, the long-term relationship proved to be mutually beneficial—or symbiotic. The bacterial cell, in short, became an asset of the archaeal cell, and the two survived together in a new way. This cell was the first eukaryote, and thus the last eukaryotic common ancestor (LECA) from which all other eukaryotes descended.

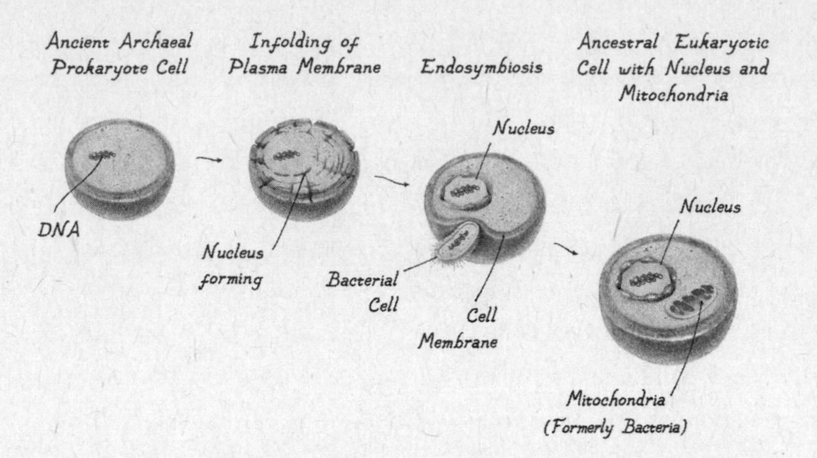


**Figure 15.1:** ***The Endosymbiotic Theory of the Origin of Eukaryotic Cells***

Just what benefit did having a bacterial partner provide for archaea? Nick Lane suggests that archaeal cells, prior to engulfing bacteria, lived on gases—hydrogen and carbon dioxide ($CO_2$). Once an archaeal cell had a bacterial cell inside it, it was able to make a living in another way. Specifically, the engulfed bacterial cell effectively became a membrane-enclosed organelle. In the process of becoming a subsidiary of the archaeal cell, the bacterial cell lost many of its own genes, but retained a small number that allowed it to physiologically maintain its cell membrane and to carry out certain select functions. This bacterial organelle is the origin of the mitochondria of eukaryotes, the membrane-enclosed energy structures we encountered in the previous chapter. The job it took on was the production of cellular energy, and because energy is necessary for all aspects of cell function, having a specialized energy machine in the cell was very advantageous.

The classic Darwinian view of evolution emphasizes divergence—new species are created by the accumulation of small changes over time that slowly transform older life-forms into new ones. The symbiotic relationship that developed between archaea and bacteria to form eukaryotes challenges this idea by emphasizing convergence—the creation of new species by the merging of existing life-forms.

The origin of eukaryotes by fusion of bacterial and archaeal cells is known as the endosymbiotic theory. Its most vocal proponent was the late biologist Lynn Margulis, who succeeded in changing the conversation about how eukaryotes came to be. While Margulis also argued that much of subsequent eukaryotic evolution (in the form of multicellular life—plants, fungi, and animals) continued on the path of convergence through symbiosis, only the more limited role of evolutionary convergence in the symbiotic joining of bacteria and archaeal cells is now widely accepted. But hers was a huge insight that had tremendous implications for understanding one of the major transitional points in the history of life.

The first eukaryotic organisms were thus single cells with internal structures—organelles—that were encased in membranes and performed important cell functions. One organelle, the nucleus, housed the vast majority of the genetic material, while another, the mitochondria, served as energy factories. These cells also had a cytoskeletal transport system for distributing chemicals throughout the cell. Unicellular founding members of the kingdom Protista are the ancestors of all complex, macroscopic forms of multicellular life. Because all of the cells that make up the bodies of complex organisms are eukaryotic, plants, fungi, and animals are themselves, as a whole, also eukaryotes.

*Chapter 16*

# BREATHING NEW LIFE INTO OLD

A fundamental characteristic of life and its many manifestations, such as behavior, is the use of carbon-based compounds to make energy. How various organisms manage to do so is fundamental to how they differ. Among eukaryotes, plants have adopted one method, and fungi and animals another. These variations are a direct consequence of what happened when an archaeal cell engulfed a bacterium.

Animals obtain energy from carbon-containing compounds by consuming and digesting other organisms (animals, plants, or fungi). Fungi also get energy from other organisms, but they don't actually consume the organism itself; rather, they begin the process by releasing digestive chemicals externally, and then consume the product. In both cases, the end result is glucose, which is delivered to cells, where the mitochondria use oxygen to make energy from by-products of glucose breakdown. This process is called cellular respiration (figure 16.1). Plants mainly make energy using chloroplasts, which capture sunlight, in a process called photosynthesis. Water absorbed from the roots and carbon dioxide extracted by leaves are broken down to yield glucose, which is stored as starch and used as fuel (food). Plants also have mitochondria and use them to make energy in darkness.

These two approaches to cellular energy production in animals and fungi on the one hand and plants on the other are derived from the way two different kinds of eukaryotes developed when archaea consumed bacteria. Some ancient bacterial cells took in oxygen and used it to break down organic compounds to make chemical energy, while others took in

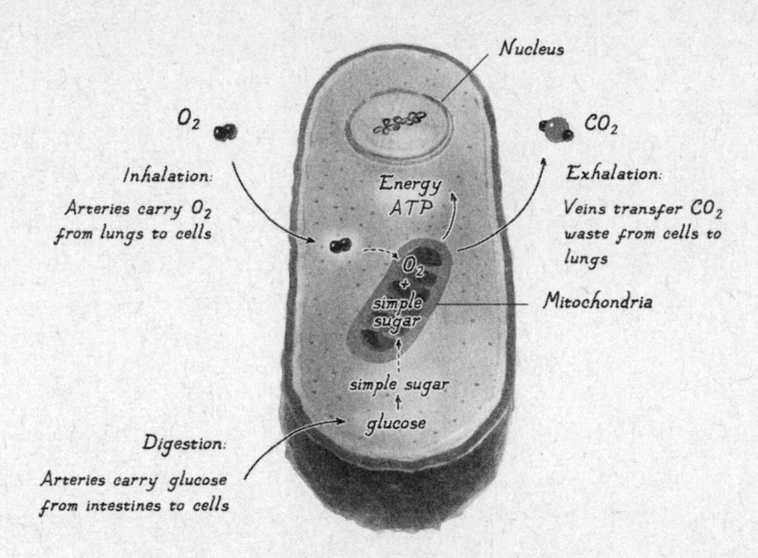


**Figure 16.1:** ***Making Energy from Cellular Respiration***

carbon dioxide and used photosynthesis to make chemical energy. Archaeal cells that swallowed oxygen-dependent bacterial cells ended up with mitochondria functioning as oxygen-dependent metabolism machines. These archaeal cells containing mitochondria then additionally swallowed photosynthesis-dependent bacteria that became chloroplasts. The chloroplasts became energy machines that use atmospheric carbon dioxide to make energy. This is how the two main categories of eukaryotic cells that spawned all macroscopic forms of life (plants, fungi, animals) arose (figure 16.2).

A particularly important factor here was the symbiotic relationship between organisms with different energy styles. The by-product of photosynthesis is oxygen. As photosynthetic-dependent microbial organisms began to multiply, which they did rapidly, the atmosphere contained for the first time sufficient amounts of oxygen to foster a growth spurt of oxygen-dependent organisms. In turn, the release of carbon dioxide by a growing number of cells with mitochondria promoted the expansion of photosynthesis-dependent organisms. The oxygen released by plants thus made life possible for animals and fungi, while carbon dioxide released

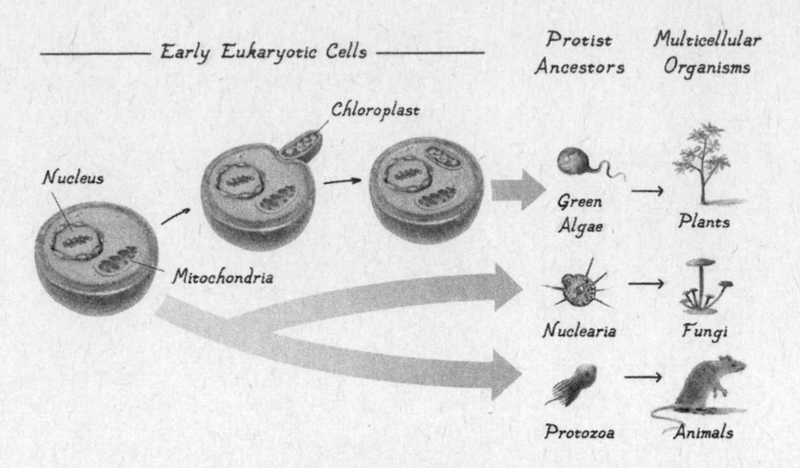


**Figure 16.2:** ***Protist Ancestry of Plants, Fungi, and Animals***

by animals and fungi made plant life possible. Tyler Volk has calculated that the recycling of carbon dioxide between oxygen-breathing and photosynthetic organisms increased global photosynthesis by two hundredfold over what it had been when $CO_2$ was supplied only by volcanoes and rock weathering.

Two kinds of souped-up archaea were thus the first eukaryotes. They belonged to the kingdom Protista, which, as noted, contains diverse unicellular organisms (paramecia, amoeba, algae, certain parasites, and other unicellular organisms). But Protista is also home to some simple multicellular life-forms called multicellular colonies. The latter are believed to have been a transition point between simple and complex life, and will be discussed later.* First, though, we need to examine in more detail two of the ways in which single-cell eukaryotes differed from prokaryotes: their larger size, and their engaging in sexual reproduction.

* Thanks to Iñaki Ruiz-Trillo for insights into protozoan ancestors of multicellular organisms.

Part 4

# The Transition to Complexity

*Chapter 17*

# SIZE MATTERS

In spite of being around for at least 3.5 billion years, prokaryotes never took the evolutionary path that would lead to a complex, multicellular, macroscopic existence. This only happened in eukaryotes. And if it had not occurred, life on Earth would have continued to be microscopic, and our planet would appear lifeless to the naked eye. A key step toward the development of macroscopic bodies was the ability of eukaryotic cells to expand in size.

The arrival of eukaryotes about 2 billion years ago disrupted peace on Earth. With their considerably larger size, they were the first true predators, and prokaryotes fell prey to them. But eukaryotes also preyed on other eukaryotes, which led to an evolutionary arms race (to use a term popularized by Richard Dawkins) between them. Being bigger helps not only in capturing prey but also in avoiding being captured. Natural selection thus selected for eukaryotes to grow larger and larger bodies.

Why, then, couldn't prokaryotes also grow bigger? Nick Lane has calculated that they had sufficient genetic raw material to make that possible, so why were they unable to turn that biological capital into a larger, more complex existence?

Biologists have come up with a number of possible answers, but the most compelling seems to be one advocated by Lane, which concerns the amount of energy that can be produced per gene. Bigger bodies require more energy, and Lane has calculated that eukaryotes can generate more than two hundred thousand times the amount of energy per gene than prokaryotes. This abundance of energy, Lane says, allowed eukaryotes to expand into bigger bodies; the inability to bump up energy supplies destined prokaryotes to microbial existence. You can probably guess why eukaryotes can make more energy, and thus grow to larger sizes—because they have mitochondria, which use oxygen to make energy in more efficient ways than alternative means.

The first major increase of oxygen in the history of the Earth occurred a bit more than 2 billion years ago, just before the arrival of eukaryotes, and, as noted earlier, is thought to have resulted, at least in part, from an increase in photosynthetic prokaryotes that release oxygen. A second rise in oxygen levels, which was significantly greater than the first, began around 800 million years ago. This additional increase helped make possible even larger, energy-demanding multicellular organisms (animals, plants, and fungi) and contributed to their diversification.

The efficiency of mitochondria as energy machines accounts for how eukaryotes were able to generate more energy per gene than prokaryotes. While some prokaryotes depended on oxygen for energy, they did not have dedicated mitochondria to make sufficient energy to grow bigger. The lesser energy-generating capacity of prokaryotes was not a problem so long as they stuck with small-volume bodies. And that's what they did—any that experimented with larger bodies would not have done well; catastrophic consequences result if cell volume increases but the energy supply can't keep up with the demand.

While size contributed positively to physiological viability and fecundity in eukaryotes, it also had its costs. It's harder to keep a big cell alive and well, as it requires more energy, more structural support, and more ways to move useful things in and wastes out. Mitochondria solved the energy problem, the cytoskeleton took care of the structural problem, and transporter proteins the carriage problem.

Lest we feel sorry for bacteria and archaea for their puny size, we should note that they have an enviable track record of survival—no other organisms can boast of having persisted for 3.5 billion years in spite of climate and other cataclysmic changes that wiped out larger eukaryotes, including many animal species. In contrast to the billions of years prokaryotes have survived, animal species on average have tended to last only about 400,000 million years before becoming extinct.

John Gerhart and Marc Kirschner argue that prokaryotes were able to survive so successfully, despite being small, because they diversified biochemically, which enabled them to adapt to environmental changes without having to change structurally. Eukaryotes, on the other hand, are specialists that changed their body structure quickly and drastically, forming the diversity of life as we know it.

*Chapter 18*

# THE SEXUAL REVOLUTION

It's one thing to suggest an explanation for *why* unicellular eukaryotes can achieve a larger mass and unicellular prokaryotes can't. But it is another altogether to account for *how* these somewhat larger (but still microscopic) cells paved the way for specialized macroscopic organisms that have many such cells that adhere together from birth, communicate with one another, and differentiate into cell types, tissues, and organs that depend on one another to survive. A key factor was that eukaryotes came up with a novel way of transferring genes to offspring. They invented sexual reproduction.

Recall that prokaryotes reproduce asexually through simple cell division. Because each daughter cell receives a complete set of genes from the mother cell, genetic diversity is not introduced asexually. But in sexual reproduction, genetic diversity is added (figure 18.1).

Sexual reproduction involves interactions between the two different mating types. What we typically refer to as "males" and "females" are complex organisms that come in two body forms that make possible the behavioral act of copulation, in which a sperm cell from the male fertilizes an egg from the female. But sexual reproduction actually began in eukaryotes long before such complex organisms existed. In other words, sex began with unicellular protists—the entire organism was essentially a free-roaming sperm or egg cell. How eggs and sperm came to be will be discussed later in the chapter, after we learn more about sexual reproduction.

The traditional scenario explaining how sex arose is that eukaryotes

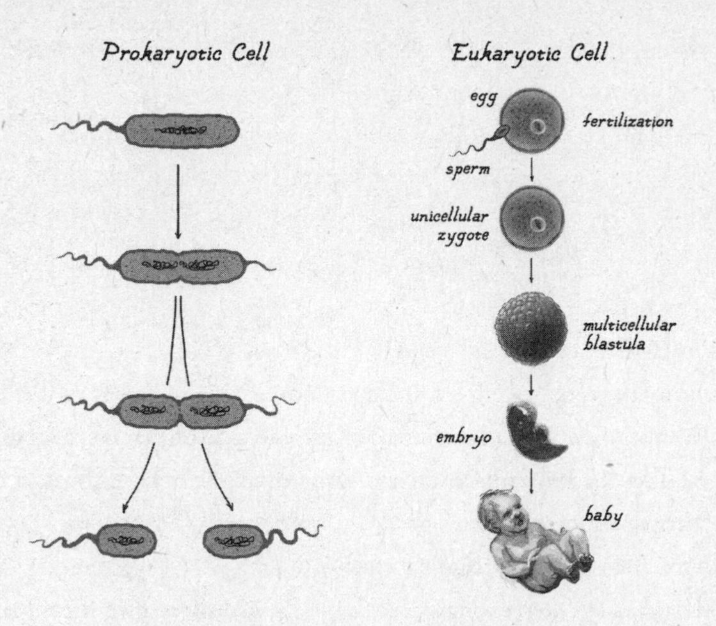


**Figure 18.1:** ***Asexual Reproduction in Prokaryotic Cells and Sexual Reproduction in Eukaryotic Cells***

started out, like their prokaryotic ancestors, reproducing asexually, and then discovered a different way of reproducing. This fit with existing evidence suggesting that some protists don't generate offspring sexually, and that even those that do reproduce sexually primarily engage in asexual reproduction. But new genetic markers of sexual reproduction indicate that the genetic capacity for sex is universal in eukaryotes, and has been present since LECA. Lack of evidence for sex in a particular protist may simply reflect the fact that it's hard to observe sex in such small creatures. But it's also possible that while some don't do the deed, their ancestors once did, and the capacity was lost over time. (There are costs as well as benefits to sexual reproduction, and it may have been more trouble than it was worth for some under their life conditions.)

When the sperm and egg meet up, they physically fuse so that the sperm can fertilize the egg—a mechanism that is true in both unicellular and multicellular eukaryotes. Because each sperm and egg inherited genes from both of its parents, fertilization of the egg adds the mélange of the male's genes to that of the female's. The genes are then mixed in a process called recombination, which results in each offspring's having a unique combination of genes that differs from those of both its parents. Depending on the mix of genes and how their expression is regulated in early life, the offspring become male or female. With each generation, the genes are shuffled again, with the result that no two organisms have exactly the same genes.

Let's dig deeper into the nature of inheritance via sexual reproduction, using humans as an example. The sperm and egg are called gametes (sex cells). Human gametes each have twenty-three chromosomes, which are combined in the fertilized egg (zygote) to give it forty-six chromosomes, twenty-three from each parent (the popular genetic screening service, 23andMe, gets its name from this fact).

The zygote is the starting point for the building of a multicellular organism. It divides in two, replicating itself so that each new cell gets the full inheritance package of forty-six chromosomes. Cell division of this

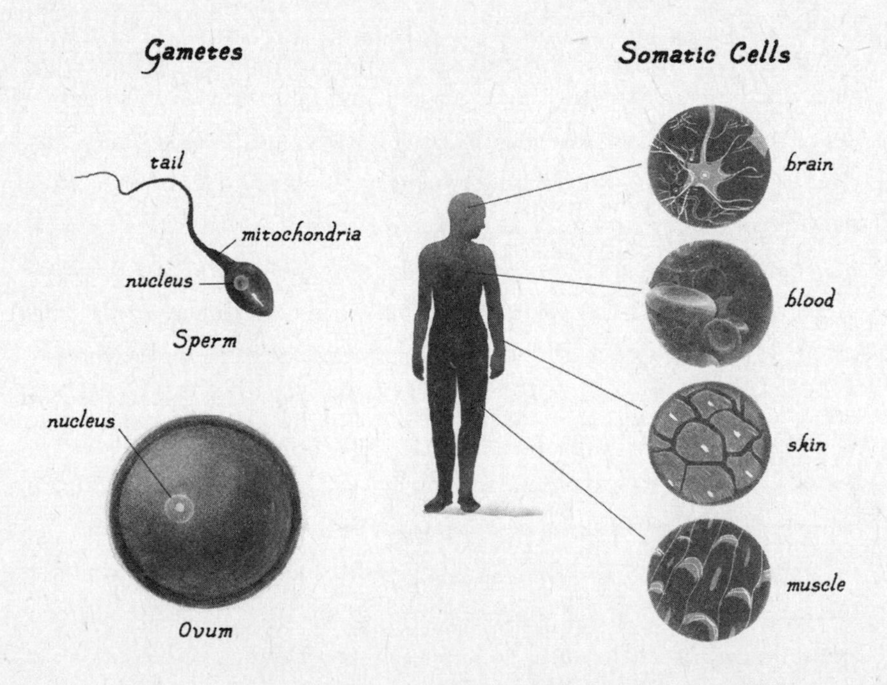


**Figure 18.2:** ***Somatic Cells and Gametes***

kind is called mitosis, and it is similar to cell division in prokaryotes. Each of these new cells further replicates and divides, and on and on. At some point, chemical signals are generated that differentiate them into the specific types of cells that make up a body's various tissues and organs (skin, heart, lung, kidney, muscle, brain, and so on). These so-called somatic cells migrate to their appropriate location in the developing organism as the body is being constructed (figure 18.2).

The other fundamental category of cells is germ cells, whose only job is to generate gametes. These selectively migrate in the body to the reproductive organs or gonads, where they remain until sexual maturity, when, through male-female interactions, the sperm can fertilize an egg.

While sex is a major source of genetic variability in eukaryotes, mutations also contribute in both useful and detrimental ways, increasing or decreasing fitness. Mutations in somatic cells affect the individual but not his or her offspring; germline mutations, on the other hand, can pass from parents to offspring (figure 18.3).

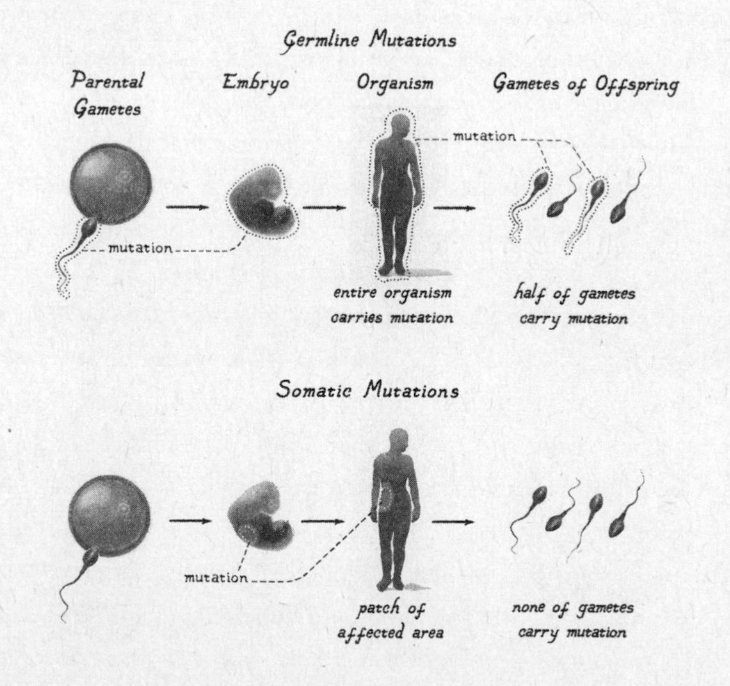


**Figure 18.3:** ***Somatic vs. Germline Mutations***

As in prokaryotes, horizontal gene transfer can also occur in eukaryotes, though to a lesser degree. One problem with GMOs is based on the very concern that genes inserted into foods might be transferred to people and alter the characteristics of the human genome.

So why did sex arise, given the fact that asexual reproduction was working pretty well for billions of years in prokaryotes? Actually, the appearance of sex is not surprising, as natural selection produces many experiments. But only those that enhance fitness become stabilized in the population over the long run. And in the case of sex, there are some high costs that had to be overcome in order for beneficial effects to lead to selection.

For example, asexual reproduction is continuous and fast (as noted above, several times per hour in bacteria). Sexual reproduction, however, is less frequent and produces many fewer offspring. Part of this is accounted for by the fact that the sperm and egg have to first find each other, and this has to happen each time reproduction occurs. Sexual reproduction is also inefficient (in complex organisms, millions of sperm are produced each day that are never used). And it is more energy demanding (complex biological processes have to be engaged to mix the genes of the two partners in the process of generating offspring).

So what *are* the advantages of sexual reproduction? The survival of a species, and even of a smaller mating group, depends not only on how well adapted the group is to the present environment, but also on how well its members can respond when their environment changes. By mixing genes from two organisms, sex adds genetic variability. A group with more such variability is more likely to have genes that will be useful under new conditions than a group with less variability. In organisms that reproduce asexually, genes (good and bad) are faithfully transmitted to offspring, and any variability comes later, in the form of mutations and horizontal gene transfer. Genetic variability encoded in germ cells and introduced into the zygote during sexual reproduction by sperm-egg interactions creates more variability in the population of reproducing organisms, and enables the population as a whole to better adapt to changing environmental conditions by increasing the possibility of useful traits. For example, in the age of global warming, polar bears that have traits that aid them in adapting to the retreat of the polar ice cap will do better than those that don't, and these traits will increase in frequency in the surviving population. Furthermore, because each new organism that results from sexual reproduction has a unique combination of genes, the deck of genes is constantly reshuffled. This means that harmful mutations that impair individual survival tend to get weeded out before they have a chance to become established in the population. (An organism that dies before reproducing doesn't contribute to the gene pool of the population.) By contrast, individuals with useful mutations that affect their germ cells and give them traits that allow them to survive long enough to mate will pass on those mutations through their offspring, and these, if sufficiently abundant in the population, will be propagated and stabilized.

As with other survival behaviors, we tend to attribute psychological significance to sex in other organisms since for us it is associated with significant psychological experiences. But the extent to which the psychological states like those we have occur in other animals is not to be taken for granted. Sex certainly didn't start off in protists as a psychologically motivated behavior. And where we draw the line between organisms that have such experiences and those that do not requires that we understand what underlies our own emotional experiences, a topic we get to later.

*Chapter 19*

# MITOCHONDRIAL EVE, JESSE JAMES, AND THE ORIGIN OF SEX

Sex is fundamental to "eukaryote-ness." It appears to have been present in the first single-cell protists, or at least very early in their history. We've examined its advantages and why it persisted. But what made it possible? Mitochondria not only hold the key to the larger size of eukaryotes relative to prokaryotes, but also to their reproductive habits.

While most DNA of a eukaryotic organism is contained in the nucleus, some is contained in the mitochondria. In sexual reproduction the nuclear genes of the two mating partners are mixed, but the mitochondrial DNA is passed mainly from one parent, usually the female, to the offspring. Both male and female offspring acquire the mitochondrial DNA of the egg, but only the female offspring can pass these genes on to her offspring. This is why every female of our species who has ever lived is said to be connected, via mitochondrial DNA, to the first woman, so-called Mitochondrial Eve.

The uniparental inheritance of mitochondrial DNA was used to prove that the infamous nineteenth-century bank robber Jesse James was indeed buried in the grave that bore his name. There had been some question about whether his death had been staged as a way for him to adopt a new identity and escape pursuit by the law. But comparison of DNA from hair and teeth in the grave with mitochondrial DNA from two living males (black boxes) that had descended from James's sister Susan and her female descendants (white circles) provided the proof (figure 19.1). The same technique has also been used to identify the remains of unknown soldiers in war.

To understand why the uniparental inheritance of mitochondrial DNA

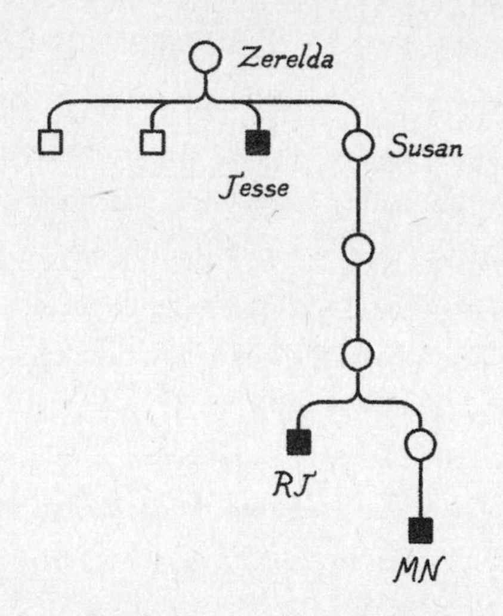


**Figure 19.1:** ***Jesse James's Remains Identified by Mitochondrial DNA of Descendants of His Sister Susan***

may have been important to the invention of sex, we have to go back to that special moment in time when an archaeal and bacterial cell entered into a symbiotic relationship such that the bacterial cell became the mitochondria of eukaryotes. Because genes control all aspects of cell function, introducing a whole new organism with its own genome (which was essentially a case of horizontal gene transfer) could have put the archaeal host in physiological conflict with the bacterial invader. Such intracellular conflict would increase free radicals (chemicals that cause "cellular stress") and could lead to DNA damage and mutations. (The reason that antioxidant-rich diets are believed to be beneficial is that they supposedly counteract the damaging effects of free radicals.) If such conflict had occurred, it would have challenged the ability of this new two-cells-in-one organism to survive and reproduce, take advantage of the benefits of the bacterial cell as a sequestered energy machine, and live together in symbiotic harmony. The fact that transmission of mitochondrial DNA was limited to the female parent, which eliminated half of the mitochondrial genes transmitted to the offspring, and thus also eliminated significant physiological conflict, made it easier for the two genomes to coexist.

Another factor relevant to the role of mitochondrial DNA in sexual reproduction is that sperm are behaviorally more active than eggs. Activity requires energy, the creation of which generates free radicals as a by-product, potentially damaging mitochondrial DNA over time. Because eggs are less active, free radical production is lower in them. Uniparental inheritance thus allowed the passing on of potentially undamaged mitochondrial DNA. This, in turn, further reduces the opportunities for physiological conflict and cellular stress.

Sex is at the heart of large, complex multicellular life. But not all sexually active eukaryotes evolved into multicellular, macroscopic organisms. What allowed that to happen?

*Chapter 20*

# COLONIAL TIMES

Living organisms can be divided into two categories: big and small. In slightly more technical terms, the small ones are unicellular organisms, while the big ones are multicellular.*

By some counts multicellular organisms arose as many as forty-six times in evolutionary history. However, this number includes instances where unicellular organisms banded together to form so-called colonies. Colonies are not true multicellular organisms, as their cells are not officially components of a single unitary body. But a certain kind of colony served as a launching pad for true multicellularity.

A colony is a group of unicellular organisms that adhere to one another (figure 20.1). They use chemical secretions to stay attached (adhesion molecules) and to communicate with one another (signaling molecules). Bac-

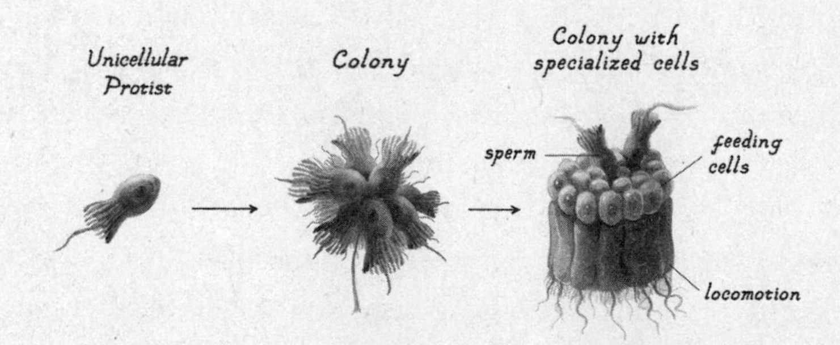


**Figure 20.1:** ***Unicellular Organisms Aggregate to Form Multicellular Colonies***

* Thanks again to Karl Niklas for consultation on multicellularity.

terial biofilm, which was described in part 2, is an example of a prokaryotic colony, but it is eukaryotic colonies that are particularly important to understanding the origins of multicellular life.

Some familiar examples of eukaryotic colonies are kelp and seaweed on the beach and green algae in ponds. Another, slime molds, which are collections of many (in some cases billions) of amoeba, have gotten attention for their huge size—one found in Texas was forty feet across. Slime molds have an uncanny ability to move across the landscape using highly efficient routes. Engineers have taken advantage of this ability by putting slime molds in specially designed mazes to help plan highway systems.

Colonies form because group existence offers advantages over unicellular life. One is safety in numbers—the large mass and density of the colony, and its ability to move as a unit, helps protect against predation. Given that true predatory organisms first emerged with eukaryotes, colonial life became especially useful as a way for prey to defend against aggressors.

Another advantage of a colonial existence is that survival chores can be divided up among different cells. When living solo, the single cell has to take care of locomotion (including approach and withdrawal), nutrient processing, and reproduction (figure 20.1). But in a colony these tasks can be spread across the cells. Since each cell has the potential to perform all functions, by suppressing the expression of some of these functions in different cells, specialization can arise. This is achieved by the release of chemicals that regulate genes that underlie different capabilities. The result is that the function that remains after all others have been suppressed is now the cell's special function in the colony.

The advantages of colonial life accrue only if all the colonists work cooperatively, so that everyone benefits. One impediment to cooperation is physiological conflict due to the genetic diversity of colony members. Defections are a problem, as they reduce the colony's size and decrease a defense against predation. One way to defect is to loaf. Loafers are moochers; they don't work (expend energy) to get resources to satisfy their needs. Defection can also occur via migration. If conditions change and it is no longer advantageous to be a member of the colony, cells can leave it and survive on their own, even if they specialized while in the colony. Since they were born with a full set of survival capacities, when the suppressing effects on gene expression exerted by other cells are removed, the lost capacities return. Defection by migration, though advantageous when conditions change, makes the defectors vulnerable to predation.

So what's the difference between colonies and true multicellular organisms? In a true multicellular organism, physiological conflict and defection are minimized by the overall genetic homogeneity of the cells. Further, division of labor is achieved by way of a genetic program that causes cells to differentiate into specific types that form tissues and organs with specific functions. Cells in different tissues come to depend on one another; they cannot survive if separated from the organism as a whole. Survival of the individual cell thus becomes subsidiary to survival of the overall organism, and mandatory cooperation results.

Colonies are the transitional step between unicellular and multicellular organisms, in that they adhere and communicate to achieve common survival goals. But there were two kinds of colonies, only one of which was able to transition to multicellular life (figure 20.2). The first consisted of genetically heterogeneous collections of individual cells that were the

**Figure 20.2:** ***Aggregating vs. Clonal Colonies***

product of cell division by different cells and that aggregated at some point in their life. In contrast, clonal colonies consisted of genetically identical cells (clones) that could all be traced back to a single mother and, after cell division, remained fused together (rather than living as individuals). Because of their genetic similarity, physiological conflict and defection were minimized.

According to Karl Niklas, the three groups that became true multicellular organisms—plants, fungi, and animals—each made the leap with the aid of a clonal colony phase of their respective single-cell protist ancestor. And each of these leaps from colonial life to true multicellularity took place via a two-step progression of natural selection, the next topic.

*Chapter 21*

# THE SELECTION TWO-STEP

One thing that never occurred to me to ask before I started writing this book is why there are only three forms of complex life (one of which, fungi, doesn't even get its own section in the grocery store). What enabled plants, fungi, and animals specifically to make the leap?*

Two key requirements for multicellular life are cell-to-cell adhesion and cell-to-cell communication. Given that these abilities are present in unicellular organisms (they are what allow colonies to form and function as a unit), what prevents colonies from becoming multicellular organisms on a routine basis? Karl Niklas says the deal breaker is the failure of most colonies to achieve the necessary sustained cooperative relationship between individual cells such that defection does not, or cannot, occur.

A number of authors, including Niklas, argue that two evolutionary steps are required to progress from unicellular to true multicellular organisms. First, an alignment-of-fitness phase has to occur in which genetic similarity among the cells prevents conflict between them, thus enhancing cooperation. Alignment of fitness is achieved by what is called the unicellular bottleneck. This means that a multicellular organism starts as a single cell, and all its other cells are generated from that individual. In animals, this cell is the zygote, the fertilized egg that contains genes from the two parents. Because clonal colonies are made up of descendants from one mother cell and thus pass through a unicellular bottleneck, they make

* Thanks to Karl Niklas for comments on the evolution of multicellularity, and export and alignment of fitness.

it halfway toward true multicellularity. But only those that advance to the export-of-fitness stage actually become multicellular organisms.

The export-of-fitness stage, the high hurdle in this race, is what is required for cells to become interdependent and to go through life with a high degree of cooperation and minimal physiological conflict. For this to happen, fitness has to be transferred from the individual cells to the organism as a whole. In the words of Richard Michod, "Units capable of independent replication before the transition only replicate as part of a larger whole after the transition." In other words, export of fitness occurs when the organism itself, rather than its component cells, becomes the unit of reproduction. When this takes place, sustained interdependence is established between cells that perform different tasks, so that cells performing one function depend on the functions performed by other cells to survive. Unlike in colonies where all cells can potentially perform all functions, in a true multicellular organism functions are programmed in the genome. Thus, defection cannot occur because cells of a given type depend on one another and are not viable alone. All this hinges on eukaryotic sex and the way an egg fertilized by a sperm becomes a new multicellular organism with features of both parents.

The greater degree of cell specificity in multicellular organisms relative to multicellular colonies, in other words, is a consequence of sexual reproduction. Previously, we discussed the difference between germ cells, which make the sperm or egg of an individual, and somatic cells, which make the rest of the cells in the body.

The idea that germ and somatic cells are different and are segregated in the body arose as a challenge to one of Darwin's key theories. He proposed that inheritance from parents to offspring involved tiny particles, called gemmules or germs, derived from cells in each body part, which aggregate in the reproductive organs and are mixed during sexual reproduction. This notion was in part inspired by Darwin's acceptance of Jean-Baptiste Lamarck's theory that traits acquired by individuals during their lives could be passed on to their offspring. August Weismann challenged this idea by proposing a theory in which cells that construct the body during early life and cells that are used later in life to transmit traits to offspring through sexual reproduction are different. In Weismann's conception, traits acquired by somatic cells, such as learned skills or mutations, are not heritable, because only the germ cells pass information between generations.

Neither Darwin nor Weismann was aware of the existence of DNA, but Weismann nevertheless was on the right track with his ideas. His theory is, in fact, still widely accepted, but with certain caveats. For example, recent research suggests that drug abuse or stress exposure in the father can predispose offspring toward vulnerability to addictive or mood-anxiety disorders. This process must occur by altering the genes of the father's sperm cells. These types of environmental effects on genes are often referred to as epigenetic influences, and provide some vindication of Lamarck.

But Eric Nestler, a leading epigenetic researcher, argues that, in contrast to genetic influences, we don't really know the extent to which epigenetic inheritance of behavior can occur. It is unclear how it would persist through the blending of genes of the two parents in the fertilized egg, in the embryonic phase, and finally into postnatal life, so as to create a particular vulnerability in the wiring of the billions of neurons in the brain. He suggests one possible scenario, called genetic imprinting, that could have such effects. In this model, one copy of a gene from a parent is permanently inhibited, setting specific vulnerabilities in motion. For example, Nestler suggests that "chronic stress might increase levels of certain microRNAs that are associated with sperm cells, which then influence gene expression in the fertilized zygote. But how that altered expression in a single cell zygote leads to altered gene expression in a specific brain circuit is completely unknown."*

We've taken this detour to a discussion of germ and somatic cells for a reason. According to Michod, the way that multicellular organisms achieve export of fitness is by sustained division of labor between repro-

---

* Thanks to Eric Nestler for comments on epigenetics. The quote was included in an email exchange on September 7, 2018.

ductive (germ) and nonreproductive (somatic) cells. Because the germ cells are physically distinct and segregated anatomically from somatic cells, the result is obligatory, nonreversible separation of reproduction from other body functions. With this division of labor the needs of the individual are sacrificed for the greater good of the organism. For example, before the evolution of sex, every cell had the capability to replicate itself. But export of fitness from the cell to the organism requires that somatic cells nonreversibly give up their reproductive rights in exchange for cooperative interdependence.

How exactly does sexual reproduction make it possible for the needs of the organism to trump the needs of its cells? Nicholas Butterfield points out that it eliminates "somatic cell parasites" (defectors), something that asexual reproduction is relatively poor at. Moreover, in sexual reproduction, harmful somatic genes are eliminated before they have a chance to stabilize. One way this happens is that some individuals with harmful genetic mutations die before reproducing. Also, in sexual reproduction, because the genes of two distinct organisms are uniquely re-sorted in each offspring, and across each generation, the genes are well mixed and shuffled, reducing the impact of any harmful ones. Finally, recall that the low activity level of the egg before the organism reaches sexual maturation reduces its chances of free radical induction of genetic mutations in mitochondria compared to cells with high energy use (i.e., sperm). The net result of these various factors is that the organism itself is favored over its component cells by natural selection in sexually reproducing multicellular organisms.

Michod gives us the big picture. Somatic cell division is essential to the viability (survival) of the organism (the cells in our tissues turn over many times during one's lifetime), but does not contribute to the ability of the organism to pass its genes on to its offspring. Germ cells, in contrast, produce gametes needed for fecundity (reproduction and genetic transmission to offspring), but contribute little to the body's ongoing viability.

In summary, the problem of sustaining cellular cooperation over the lifespan of a multicellular organism is solved by two-step selection. First, the unicellular bottleneck ensures the genetic homogeneity of the organism and minimizes physiological conflict, which in turn minimizes defection by freeloading or by migration. This is the alignment-of-fitness step. But in addition, cells in a multicellular organism belong to specific tissues with different functions, and survival of the organism, as well as that of its individual cells, depends on the distribution of labor across the tissues. For example, cardiac cells in heart muscles enable the heart to pump blood. Blood is made up of its own cells, and is pumped through arteries, made up of still different cells. Blood carries oxygen to the various tissues, where the mitochondria in each of the cells use the oxygen to make energy. The oxygen comes from air filtered by the cells of lung tissues and is used together with glucose resulting from digestive cells breaking down food. If any of these cells are separated from the organism, they die without the support provided by the other cells in other tissues. And if one organ system breaks down the other systems suffer.

*Chapter 22*

# FLAGELLATING THROUGH THE BOTTLENECK

Of the three groups of organisms that achieved true multicellularity, the ones we care about most in this story are animals, which seem to have first appeared in rudimentary form around 800 million years ago. The leading explanation for how they came about is the colonial flagellate hypothesis. Originally proposed by Ernst Haeckel in the nineteenth century, the idea was later discredited due to the lack of evidence. But modern research strongly supports Haeckel's basic notion.

Plants, fungi, and animals each have a protist ancestor (figure 22.1). The protist ancestor of animals is an ancient extinct protozoan that is also believed to be the ancestor of a group of present-day protozoa called choanoflagellates. Because animals and choanoflagellates share a common

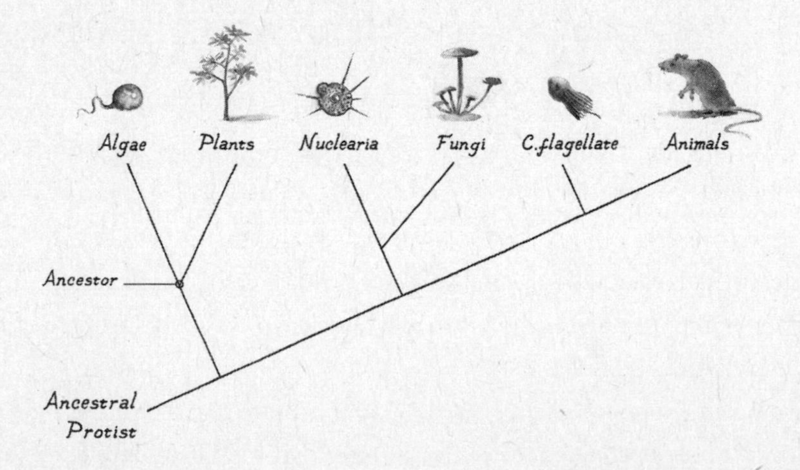


**Figure 22.1:** ***Protozoan Relatives of Plants, Fungi, and Animals***

ancestor, they are sister groups, and features shared by the two groups today provide important clues to the nature of their common ancestor, and thus help trace the deep history of animals.

Choanoflagellates get their name from their flagellum, a tail that they use to propel swimming movements (figure 22.2). Flagella movements, called beats, involve wavelike undulations. By contrast, some other eukaryotes have cilia, which are shorter and more numerous and rotate rather than undulate. While flagella are important in the transition to animals, as we will see later, cilia also played a key role in the evolution of nervous systems in early animals.

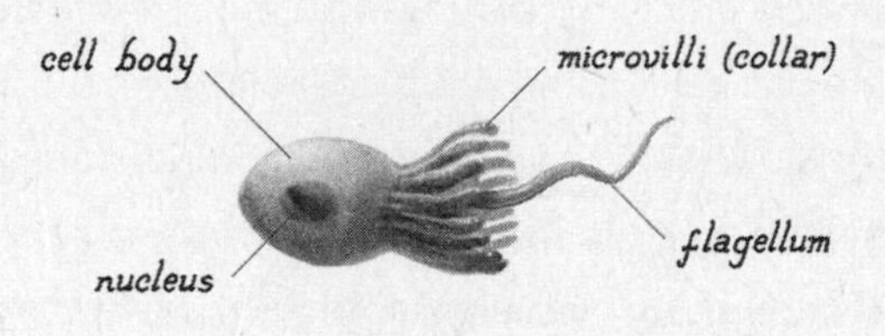


**Figure 22.2:** ***Choanoflagellates—The Closest Single-Cell Relative of Animals***

Choanoflagellates are predators, feeding off bacteria. While the flagella of bacteria enable them to move, choanoflagellates have more control in their motion. The flagella of bacteria spin in a fixed way, making approach and withdrawal possible by modulating random activity. Choanoflagellates can control the direction of their movements toward or away from nutrients or threats with their flagellum by generating electrical signals that cause beating movements.

The flagellum is connected to the cell body by a collar made up of membranes. Choanoflagellates feed by moving their flagellum to create water currents, pulling bacteria toward them and trapping them with the collar. The bacteria are then absorbed from the collar into the cell. The collar is important because, as we will later see, it is a link between choanoflagellates and the first animals, sponges.

Nicole King and colleagues have studied biological and behavioral aspects of sex in choanoflagellates, showing that they reproduce both sexually and asexually. Asexual reproduction is the norm, but under certain conditions, sexual reproduction is triggered. For example, when food supply is limited, cell survival is challenged, and asexually produced offspring that are either smaller or larger than usual are generated. These become gametes, sex cells; the small ones are sperm and larger ones eggs. When the sperm encounters an egg, they fuse, and the egg is fertilized; progeny receive genes from the two parents. Infection by bacterial parasites can likewise trigger sexual reproduction.

Key to the significance of choanoflagellates is their tendency to form colonies. For example, choanoflagellates belonging to the species *Salpingoeca rosetta* undergo asexual cell division from a single mother and form clonal colonies with genetically homogenous cells. As noted earlier, clonal colonies helped make multicellular life possible by virtue of being genetically homogeneous, minimizing cell-to-cell conflict and thus defections. As in any colony, nearby cells stick together by way of adhesion chemicals that also establish molecular bridges between them and enable cell communication via signaling molecules. The particular adhesion and signaling molecules in question were once thought to exist only in animals. The fact that genes underlying these and a number of other molecules once believed to occur only in animals have been found in choanoflagellates is part of the compelling body of evidence supporting the link between choanoflagellates and animals.

The simplest choanoflagellate colony consists of individual cells that are bound together in the form of a sphere. A slightly more sophisticated architecture involves cells forming a ring around a hollow sphere, with their flagella extending outward. When the cell bodies release certain chemicals into the sphere, the flagella move in a coordinated beating fashion, steering the mass of cells toward useful and away from harmful substances that are encountered. Nutrients extracted by cells in one part of the colony are chemically transported to their neighbors and shared via the adhesion bridges.

One problem with this simple arrangement is that a cell can't both feed and divide at the same time. And since these organisms were predators, surviving on bacteria, they had to actively feed to stay alive. Cell specialization was the response to overcome this conflict—some cells, through regulated gene expression, took over reproduction, leaving feeding and locomotion to others. As described above, under certain conditions, sexual reproduction is favored, and cell division produces either male or female mating types. The females move into the inner sphere. Males then swim into the sphere, and when they encounter females, they fertilize them. Progeny then move to the exterior and contribute to the other routine functions of the colony.

Multicellular colonies possess the basic characteristic of multicellular organisms, such as cell-to-cell adhesion, cell-to-cell communication, and division of labor between cells by control of gene expression. Clonal colonies, such as choanoflagellates, add genetic similarity, which reduces physiological conflict and minimizes defections. While they thus make it over the first hurdle of multicellular selection (alignment of fitness), they do not make it over the second (export of fitness). They do not reach the point where survival and reproduction are exported to the level of a multicellular organism, creating sustained interdependence between cells that perform different tasks. They got part of the way by using sexual reproduction to generate male and female offspring, but when these mate the end result is just additional individual choanoflagellate cells that at most can become part of a colony. By contrast, while the initial outcome of reproduction in complex multicellular organisms is also just another cell, it is a unique *kind* of cell—one with the genetic wherewithal to generate an entire multicellular organism with all its interacting parts.

Karl Niklas has noted that the evolution of true multicellular organisms occurred in small steps: life didn't just jump straight from protozoans to full-fledged animals in a single huge leap. In this regard, one important feature of choanoflagellates that anticipated later developments in animals is that they possess many of the physiological, genetic, and molecular foundations of neurons and the nervous system, the prized possession of animals. Ancestral choanoflagellates used electrical signaling to control flagella beating, which is similar to the contraction of muscles in animals. They also used electrical signals to communicate within the cell, which is a key feature of animal neurons. In addition, they have the genes, and the proteins instructed by these genes, that animals use to form synapses, key to communication between neurons. The first animals inherited these features from the common ancestor they shared with choanoflagellates. While sponges, which are believed to have been the first animals, were unable to put the pieces together to build a nervous system, they retained the relevant traits, which enabled subsequent animals to put them to use in the first nervous systems.

PART 5

# . . . And Then Animals Invented Neurons

*Chapter 23*

# WHAT IS AN ANIMAL?

The formal name of the animal kingdom is Metazoa (meaning "about animals"). Animals are defined biologically by how they differ from organisms in the other two multicellular kingdoms—an animal, in other words, is a multicellular organism that is not a plant or fungus. As previously noted, these groups each evolved from a different unicellular protist ancestor. Each survived in somewhat different ways and passed on somewhat different genes to their offspring. These genes, in turn, gave rise to unique bodies with distinct solutions to the problems of surviving and reproducing. While many differences resulted from the unique evolutionary paths that separated the different types of multicellular organisms, three are especially important.

One is how each group ended up going about energy management: As we've seen, animals get their energy by consuming and digesting other organisms, including plants or fungi, or even other animals, while fungi digest other organisms externally and then absorb the digested product. Plants for the most part go it alone—they make their own energy through photosynthesis rather than from eating or absorbing other organisms. A second striking difference is that animals are motile: They move around from place to place in their world. Many can respond with precision to sudden changes in environmental conditions. This ability was especially useful in acquiring food by enabling them to capture and consume motile prey, but it also helped in escaping from their own predators—it's easy to see why predator-prey interactions have driven many aspects of animal evolution. Rapid and precise movements were made possible by a third difference: Only animals evolved nervous systems and muscles, which greatly extended their behavioral options.

The evolutionary path from protozoa to multicellular animals is depicted in figure 23.1. With more cells, more energy was required. And this required oxygen. It is thus of interest that a dramatic rise in concentration of oxygen in the atmosphere about 800 million years ago coincides with earliest evidence for animal life.

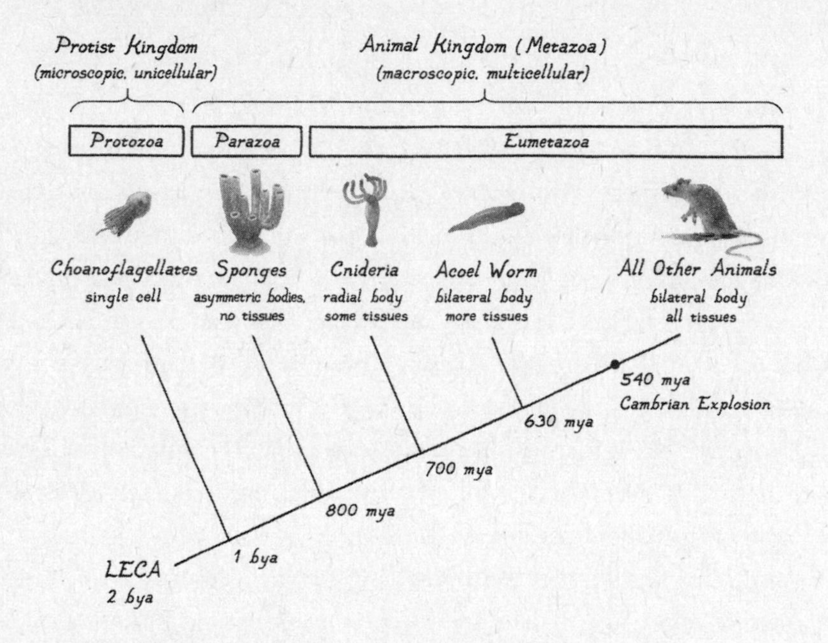


**Figure 23.1:** ***Transition from Protozoa to Animals (Metazoa)***

So which creature was the first animal? The term *protozoa* literally means "first animals," but this designation is held over from a time before animals came to be defined in terms of multicellularity. These days, the generally accepted view, which I will follow here, is that sponges, which belong to phylum Porifera, were first. While the earliest fossils of possible sponges date from about 650 million years ago, their arrival date has been pushed back to about 800 million years ago, based on molecular evidence. Their absence in the fossil record for 150 million years reflects two factors. One is that geological and atmospheric conditions at the time were, in general, not conducive to fossilization, but even if more favorable conditions had existed, fossilization would have been limited because sponges have soft bodies that do not fossilize well.

Though sponges have multiple cells, they lack tissues, and thus were not originally considered part of the animal kingdom for some time. They occupied a kind of intermediate status—more than a protozoa but less than a true animal. However, the discovery that they have genes that are otherwise found only in animals helped cement their place in the metazoan category. Lacking tissues, though, they are still somewhat in between Protozoa and true animals, and for this reason, as we will see in the next chapter, are an important link between unicellular Protozoa and the rest of the animal kingdom.

Sponges are part of a subkingdom of metazoan life called Parazoa (literally, "next to animals"), which only includes one other group. Placazoa are the other members of Parazoa. Like sponges, they are tissueless multicellular animals. While there are many varieties of sponges today, trichoplax is the only known placazoan in existence.

All animals with tissues that form organs and systems belong to the metazoan subkingdom Eumetazoa ("true animals") (table 23.1). Two early eumetazoan phyla were Ctenophora (comb jellies) and Cnidaria (hydra, jellyfish, sea anemone, and corals). Parazoa, together with Ctenophora and Cnidaria, are considered basal metazoans, animals near the base of the animal tree of life (figure 23.2).

TABLE 23.1: Animal (Metazoan) Subkingdoms

**Parazoa**—a subkingdom of Metazoa with only a few differentiated cells and with minimal tissues. The key phylum is Porifera (sponges).

**Eumetazoa**—a subkingdom that includes animals with differentiated cells that form tissues, organs, and systems. Most animals that are not sponges are eumetazoans, including all radially symmetric (cnidarians like hydra and jellyfish, and ctenophores like comb jellies) and bilaterally symmetric (all other invertebrate and all vertebrate) animals.

**Figure 23.2: *Basal Metazoans***

Above I said that the conventional view is that sponges were the first animals. While there has actually been considerable debate about whether Ctenophora or sponges came first (figure 23.3), the sponge-first scenario is more widely accepted, and we proceed assuming this to be the case.*

It is often said that these basal metazoans have not changed much over the past half billion years or so. But the buildup to the fairly stable body plans that now exist in these phyla likely involved quite a bit of experimentation between the time these various groups initially parted ways and when they each stabilized. For this reason, Allen Collins, a leader in research on the evolutionary history of early metazoans, urges caution when talking about basal metazoans in terms of present-day examples, since today's sponges and jellyfish are closely related to their extinct ancestors,

* In the sponge-first scenario, sponges diverged from the protist ancestor, and Ctenophora and Cnidaria are sister groups that separately evolved from sponges. In the Ctenophora-first scenario, Ctenophora and sponges separately diverged from protist ancestors, and Ctenophora are a sister group to all other animals. Since Ctenophora have nervous systems, this scenario would require sponges to have lost their nervous system, and then Cnidaria, in evolving from sponges, would have to have reacquired one. Since traits are lost and regained throughout evolutionary history, this is not damaging to the Ctenophora-first hypothesis.

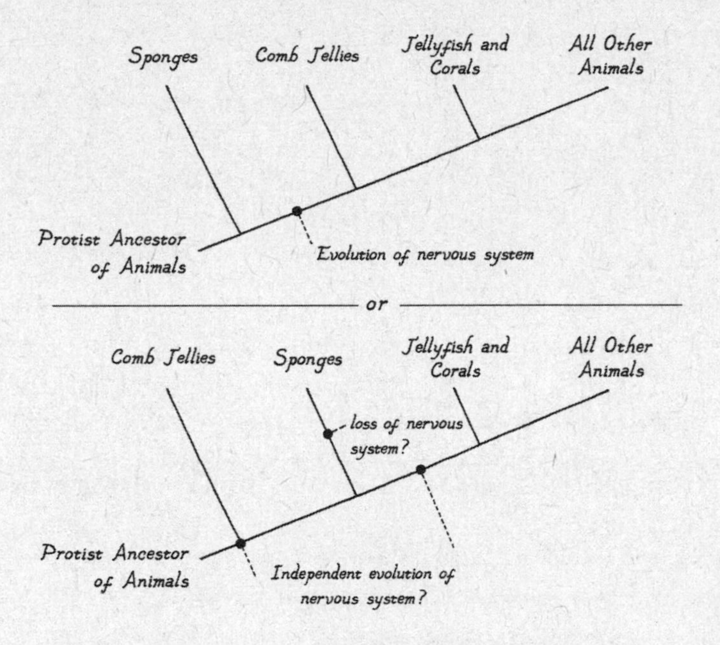


**Figure 23.3:** ***Two Scenarios of Origins of Basal Metazoans and Their Nervous Systems***

but are not identical. What has remained unchanged is their relatively simple basic body designs.*

Animals are grouped together as members of a species, and species are grouped together within phyla, because their members have body designs that are more similar to one another than to the bodies of animals in other groups. The body design of an animal is often referred to by the German word *Bauplan.* Many millions of species of animals, each with a unique Bauplan, exist, but all these various body plans fall into three general categories: asymmetric, radial, and bilateral (table 23.2).

Parazoa (sponges) have irregular shapes and are typically said to have asymmetric bodies. Cnidaria and Ctenophora have definite shapes and can be sliced into two equal parts. In contrast to a sponge, slicing the body

* This caution is applicable to the relation between all present-day organisms and extinct ancient ancestors.

TABLE 23.2: Body Plans in the Animal Kingdom

| ASYMMETRIC BODY PLAN |
|---|
| Porifera (sponges) |
| **RADIAL BODY PLAN** |
| Ctenophora (comb jellies) |
| Cnidaria (hydra, corals, anemone, jellyfish) |
| **BILATERAL BODY PLAN** |
| Invertebrates (worms, arthropods, mollusks, starfish, lancelets) |
| Vertebrates (fish, amphibians, reptiles, birds, mammals) |

of a jellyfish in half by cutting through it from top to bottom yields symmetric halves no matter where on the circular umbrella-like top you make the cut. Such shapes are said to be radially symmetric (symmetric around the radius of the circle) (figure 23.4).

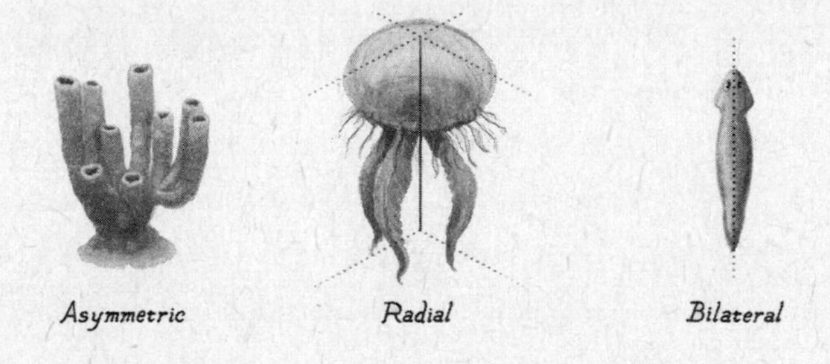


**Figure 23.4:** ***Asymmetric, Radial, and Bilateral Body Plans***

Around 630 million years ago nature's bodybuilding business took a third direction—a worm appeared with a novel body plan. Like a jellyfish, it had a top and bottom, but in addition it had a definitive front and back. The only way to cut such an organism into symmetric halves is through the long (front-to-back) axis. When this is done you end up with complementary left and right sides. Such an organism is bilaterally symmetric. Today, more than 99 percent of all the animals on Earth are bilaterally symmetric. Most if not all of these are descendants of this ancestral worm, which is known as the last common bilateral ancestor (LCBA), and is believed to have been an acoel, a tiny marine flatwormlike creature.*

Once the acoel worm acquired a bilateral body plan, it didn't take long for bilaterals to diversify into the major groupings that now exist. Much of the excitement started around 540 million years ago, and was over by 480 million years ago. This period of rapid expansion of the bilaterals is called the Cambrian Explosion (see figure 23.1), by the end of which the major phyla of today were all in place (figure 23.5). This is not to say that all the various animal species that exist today came into existence during the Cambrian Period (the phyla were defined but many of our contemporary species came later). The animals of the Cambrian time were all sea-

*1. Pikaia (cephalochordate) 2. Jellyfish (cnidaria) 3. Hydra (cnidaria) 4. Crinoids (echinoderm) 5. Comb Jellies (ctenophora) 6. Haikouella (vertebrate) 7. Trilobite (arthropod) 8. Sponges (porifera) 9. Worms (priapulida) 10. Conodont (vertebrate) 11. Anomalocaris (arthropod) 12. Nautilus (mollusk) 13. Anemone (arthropod)*

**Figure 23.5: *Cambrian Explosion***

* It's hard to obtain conclusive evidence about the exact nature of ancient animals. There is some question, for example, about whether sponges might have begun as bilateral organisms and then lost this capacity, and even whether they might once have had tissues, including neurons, and later lost them. Since there are no firm answers we stick with the conventional view as we proceed.

dwelling creatures, as the invasion of the land had not yet taken place. Above-sea-level mayhem came a bit later.

We'll have much to say about bilaterals and their various expansions. But for now our goal is to explore the basal metazoans, and specifically, to see how asymmetric, tissueless sponges arose from a unicellular protozoan ancestor, and paved the way for radially symmetric cnidarians with tissues, which in turn were precursors to bilaterals.

*Chapter 24*

# A HUMBLE BEGINNING

Sponges not only lack tissues and a well-defined body shape but are also unimpressive in the behavioral category, being largely immobile, sessile creatures that spend most of their adult lives attached to a fixed location.* Their soft bodies have an inner cavity with an opening (mouth) called the osculum at one end; the other end (the base) is usually attached to a substrate (figure 24.1).

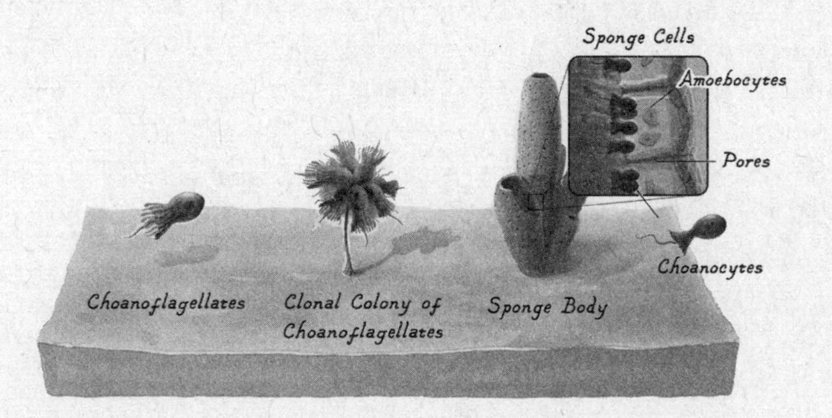


**Figure 24.1:** ***Choanocyte Cells Link Sponges to Single-Cell Choanoflagellates***

Despite their lack of true tissues, organs, and systems, sponge bodies are made up of different types of cells, one of which is particularly important for understanding the evolution of multicellularity. This is the cho-

---

* Maja Adamska provided me with useful information on sponge physiology and behavior.

anocyte cell. These cells have a flagellum and collar, and are strikingly similar to unicellular choanoflagellates. It's as if the sponge body was built from a colony of them with a stalk attached to substrate (see figure 24.1). Choanocytes reflect the genetic endowment handed down from their flagellated protozoan ancestors, but are not simply unicellular flagellates that cling together.

In the process of becoming true multicellular organisms, the flagellated ancestor went through an alignment-of-fitness and export-of-fitness phase that enabled some of its features to be programmed into the genome of what became multicellular sponges, allowing them to reproduce by generating a zygote, a cell from which other cells can differentiate into a complete multicellular organism. According to Carl Nielsen, an advanced clonal colony stage consisting of choanocytes originating from one cell (a fertilized egg), clinging together and sharing nutrients with neighboring cells, was key. This organism, Nielsen argues, was the first metazoan, the ancestor of present-day sponges, and of all other animals that came afterward.

Choanocyte cells line the inner cavity of the sponge body. Its outer surface or epidermis is formed by a layer of cells called pinacocytes, which protects its body, much like our skin. Unlike human skin, however, the pinacocyte cells are not tightly connected to one another, hence their disqualification as a kind of true skin tissue. Some sponges also develop an outer shell made of calcium carbonate, but neither is this tissue, since it is a mere physical, not a biological, structure—a shell made of inorganic chemicals rather than living cells.

The space between the outer surface and inner cavity is called the mesophyll, and contains an endoskeleton (inner skeleton) made up of a squishy material called spongin, the stuff from which household sponges were traditionally made (today, most are synthetic). The endoskeleton of some sponges also has mineralized particles of calcium carbonate that add support.

Given their stationary nature, sponges defend themselves against harm not by fleeing but by releasing noxious chemicals into their surroundings, and with external calcified barbs that deter consumption. Nevertheless, sponges do depend on active, behavior-like activities for their survival. Like all animals, they require nutrition. Sally Leys and others have shown how sponges acquire food by moving water through their inner cavity (figure 24.2). This is a primitive form of feeding behavior.

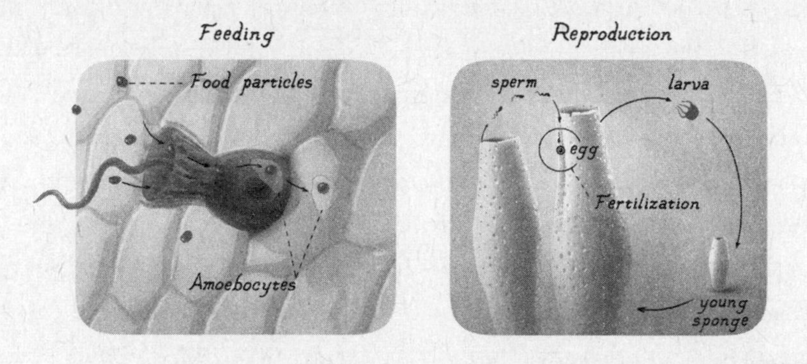


**Figure 24.2:** ***Feeding and Reproduction in Sponges***

The beating of choanocyte flagella in the inner body pulls water in through pores in the pinacocyte layer, and keeps it flowing continuously through the cavity, upward toward the mouth. As the water passes through the pores, large particles are filtered out, while smaller food items, especially bacteria—sponges' main source of food—enter. The bacteria are captured by the choanocyte collars in the mesophyll, where they are absorbed and stored in packages composed of amoebocyte cells. Adjacent cells feed off these packages, but the amoebocytes are also mobile, and they move through the mesophyll and distribute the nutrient packages to other cells in the sponge's body.

Wastes ($CO_2$ and ammonia) and sediments (sand) are carried out by water flow. But in addition, sponges use a whole-body movement, comparable to a sneeze, to expel the unwanted contents, essentially cleaning out the cavity. First the body inflates, and then collapses, moving water out. The inflation is controlled by myocytes, a term that is usually applied to muscle cells, though the myocytes of sponges are not true muscle tissue, and are not controlled by nerves, since sponges do not have neurons. Myocytes are best thought of as contractile cells that are precursors to muscles. The contractions underlying the sneeze are wavelike and are controlled by chemicals that diffuse throughout the body and are thus slow, more like the slow peristaltic actions of our digestive muscles than the fast action of our skeletal muscles.

Like choanoflagellates, sponges reproduce both asexually and sexually. Sexual reproduction can occur when choanocytes detach from the inside of the sponge cavity and transform into sperm cells (spermocytes) (see figure 24.2). It's no accident that sperm resemble choanocytes. Some sponges are hermaphrodites, and self-fertilize their eggs. Sexual reproduction can also occur between individual male and female sponges, the result of which is the mixing of genes from the sperm and egg, and thus the programming of a new, individual, multicellular sponge with a unique genome.

Sponges differ in an important way from both choanoflagellates and later eumetazoans—they are minimally motile. However, during their larval phase early in life, this is not the case. And, as we will see later, free-swimming sponge larva played an important role in the evolutionary path to the rest of animal life.

Choanoflagellates overcame the alignment-of-fitness barrier to multicellularity. But sponges did what choanoflagellates and their ancestors were unable to do—overcome both the alignment- and export-of-fitness barriers. They start life as a single cell that produces a complete, whole organism. Still, unlike in most later animals, their germline is not sequestered—many cells in the sponge can reproduce the entire body.* Despite this, they did pave the way for future animals to possess two important characteristics: a sequestered germline and a nervous system.

* Thanks to Sarah Barfield for comments on germline segregation.

*Chapter 25*

# ANIMALS TAKE SHAPE

How did complex animals with tissues, organs, and systems emerge from the amorphous shape and tissueless bodies of sponges? Why did it only happen once? And why did it happen by a choanoflagellate-like protozoan becoming a sponge, a sponge becoming cnidarian, a cnidarian becoming a bilateral invertebrate, and a bilateral invertebrate becoming a vertebrate? Thomas Cavalier-Smith, a leading evolutionary biologist, hypothesized that a key step in this progression to complex animals and their nervous systems was the evolution of a novel way of managing nutrition.

The eukaryotic protozoan choanoflagellate-like ancestor of animals was a predator that fed off small bacterial cells. But being a single cell, it did not have a mouth or digestive system. Instead, it made energy by absorbing food. Sponges do the same, but use many cells in the process. As we've seen, two kinds of cells are involved—choanocytes capture, and amoebocytes absorb and transport. But eumetazoan animals that followed had cells that formed digestive tissues, organs, and systems. For example, the cnidarian descendants of sponges had a specialized mouth organ connected to a digestive system.

Cavalier-Smith argues that sponges are the only animals imaginable that could have paved the way to this more complex approach to digestion. They evolved multicellularity without changing the absorptive mode of feeding they inherited from protozoan ancestors, but then some sponges acquired a new body plan with a new approach to predation that paved the way to a new kind of digestion (figure 25.1). These particular sponges evolved specialized cells that formed appendages (tentacles) with sharp barbs that could catch food and transfer it, through the mouth and into a specialized internal digestive organ—a gut, which was formed by sealing the body pores. Food then entered through the mouth, and was digested in the new structure. This renegade sponge was the ancestor of present-day Cnidaria, and its novel food acquisition and consumption skills paved the way for all animals that followed, including humans.

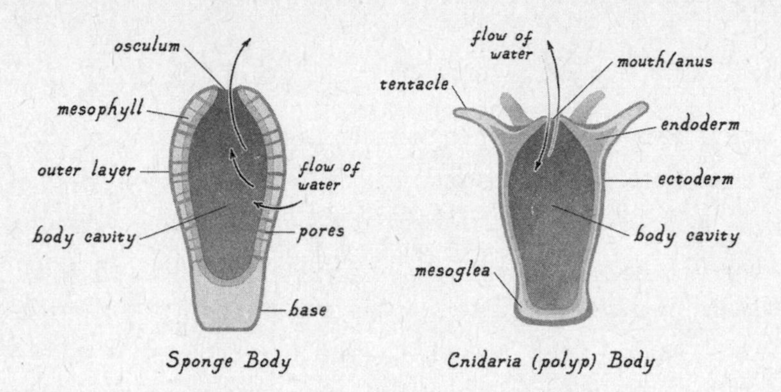


**Figure 25.1:** ***From Sponge to Cnidaria***

Cnidaria are a large, diverse group of creatures that include corals, sea anemones, jellyfish, and hydra. Jellyfish go through a life cycle in which they take different shapes. As they transition from larva to mature animals, they acquire a vaselike polyp body, similar to that of a sponge. Somewhat later, they transform into a medusa shape that resembles an umbrella (figure 25.2). The other Cnidaria, though, retain the polyp shape throughout their adult lives.

Like all eumetazoans, cnidarians are defined by the presence of tissues. Early in life, their embryos are organized in layers that generate cells destined to become tissues that compose different body parts. Cnidarians typically have two such layers. The cells in the ectoderm form the outside of the body, while those of the endoderm generate cells that form the gut. These two kinds of tissues can be thought of as forming two bags, an outer one that forms the skin that surrounds the entire body and protects it, and an inner one that forms the gut and processes food and waste. The inner

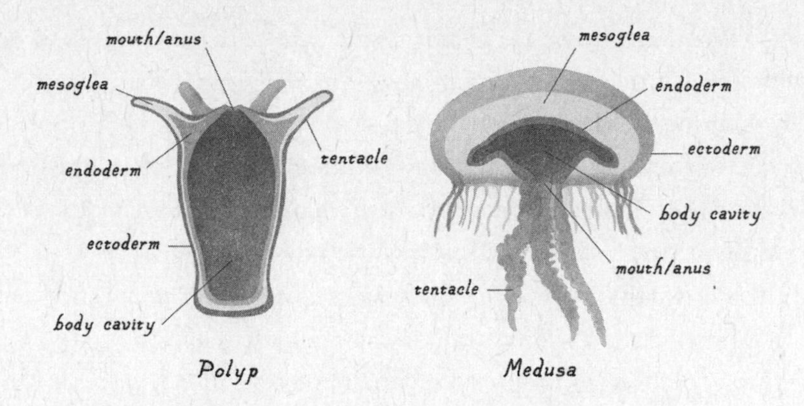


**Figure 25.2:** ***Two Cnidarian Body Types***

bag has an oral opening, which is called the mouth, but it serves as both mouth and anus.

The space between the skin and gut is filled with a gelatinous nontissue substance called mesoglea, which give "jellyfish" their moniker. While all Cnidaria have a gelatinous interior, in some, like hydra, it is more like a thin layer of glue that connects the ectoderm and endoderm than an expansive filling in and of itself. Even corals, which look more like underwater rocks than animals because of their mineralized calcium carbonate outer skeleton, have mesoglea between their ectoderm and endoderm.

Cnidaria have an array of sensory capacities. Jellyfish, for example, can respond to touch, gravity, chemicals, and light. Their photosensitive capacities are especially impressive, with some having as many as sixteen eyes positioned around the side of the umbrella, providing 360 degrees of vision.

The behavioral capacities of adult Cnidaria exceed the taxic behaviors of single-cell microbes, and also those sponges that are fairly sessile as adults. For example, Ralph Greenspan describes two kinds of swimming behaviors in jellyfish. Slow swimming is used to hunt for prey. Rhythmic contractions of the umbrella move them from place to place. They then stop, and slowly float downward in search of food. Fast swimming, on the other hand, is used to propel their bodies away from predators. If the jellyfish detects physical contact with its umbrella, the structure rhythmically contracts, but much more rapidly and powerfully than when engaged in slow hunting. This causes the whole body to jet forward.

Although jellyfish are pretty good swimmers, they are no match for the power of tides, which carry them out to sea. When water conditions are more permissive, they make their way back to the preferred shallows, using the trees or other shoreline landmarks as a visual guide. (They don't really see objects per se, but merely detect gradations of light.)

Tentacles are a key feature of cnidarian body. During the descent phase of slow swimming, jellyfish search for food by spreading the tentacles and using them as a fishing net. The tentacles are home to cnidocyte cells, from which the phylum's name is derived. When the tentacle makes contact with an object, these cells are activated, but only if the chemistry of the object is indicative of a potential prey. If so, a barb unfolds and attaches to the prey, injecting a toxic chemical to immobilize it. The tentacle then delivers the immobilized prey through the oral opening and into the gut, where gastrodermal cells release chemicals that digest the food and make energy. This behavior in Cnidaria is how the animal mode of feeding got its start.

Cnidaria reproduce sexually and asexually. Tentacles are used in sexual reproduction: the male attaches one of his tentacles to a tentacle belonging to a female, and hands off sperm to her, and she completes the fertilization of the egg. Cnidaria also engage in hermaphroditic reproduction through self-fertilization. Regardless of how they mate, the result is a single cell, a fertilized egg that gives rise to a complex, multicellular organism. Thus, like sponges, Cnidaria broke through the export-of-fitness barrier to create offspring from a single cell. But they went further, evolving a sequestered germline, which protects the individual from passing on somatic cell mutations.

Germline sequestration became especially important later, as body size increased in metazoans. Larger bodies require a greater number of somatic cell divisions during early development. With each round of cell division, mutations can enter. If the germline is segregated, somatic mutations will not accumulate and will not be passed on to offspring. Without germline sequestration by Cnidaria, the variety of life on Earth today would probably be quite different.

The behavioral repertoire of Cnidarians is more elaborate than that of sponges. Complex movements of the tentacles and body require tissues that can rapidly respond to sensory information when engaging in survival activities. Muscle tissue makes such movements possible. But to take advantage of this feature, something faster than the slow diffusion of chemicals throughout the body, such as in sponges, is required. Cnidaria developed neurons for this purpose. In fact, neurons and muscles evolved in tandem in Cnidaria. Let's see how this happened.

*Chapter 26*

# THE MAGIC OF NEURONS

As animals became more complex, consisting of multiple cell types organized into systems, new challenges were faced in terms of maintaining the integrity of the organism as a self-sustaining unit in which the parts sacrifice their individuality to maintain the physiological viability of the whole. The nervous system was the solution. A key part of our story is thus how nervous systems came to be.

Nervous systems consist of specialized cells, neurons, that can rapidly communicate across long distances. The sponge ancestor of Cnidaria didn't have neurons, much less a nervous system, but Cnidaria had both. In the next chapter, we will examine how this transition occurred, but first it will be useful to review some basic facts about what neurons are and how they achieve their magic in solving the problem of communicating between spacially separated cells by minimizing time as a factor.

Like all cells, neurons have a cell body, but additionally have nerve fibers (figure 26.1). One of these is the axon, which extends out of the cell body and enables the cell body to send messages over long distances to other neurons. Neurons typically have only one axon, but possess many of the second kind of fiber, dendrites. These extend out from the cell body like antennae for relatively short distances and receive messages transmitted from the axons of other neurons. Axons also make connections with other parts of neurons, but we'll focus on dendritic ones.

The inputs to the dendrites of a neuron from the axon of others help generate an electrical response in the cell body of the receiving neuron.

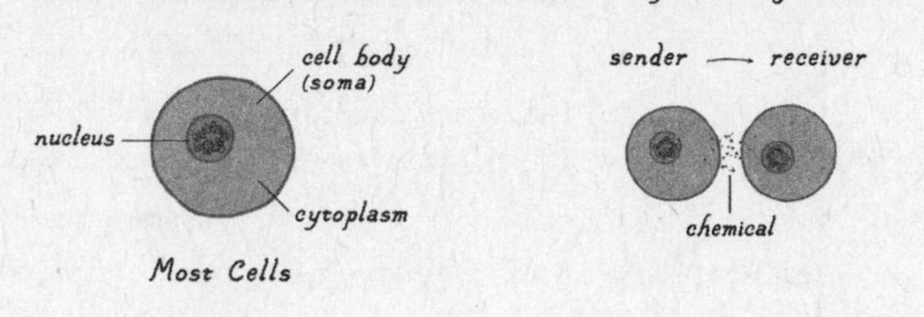


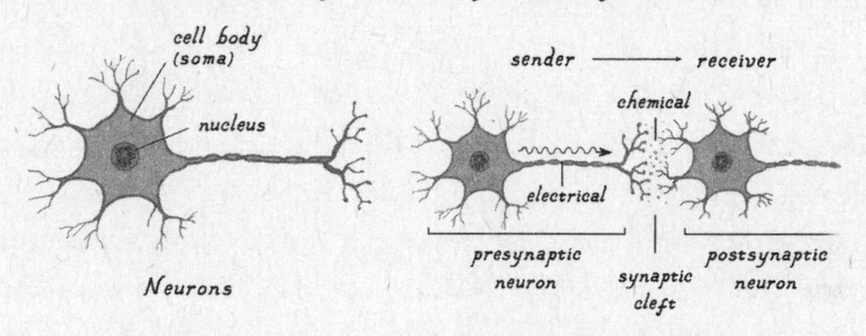


**Figure 26.1:** ***Structure of Neurons vs. Other Cells***

This action potential rapidly shoots down to the end of the axon, the terminal, and causes the release of stored packets of chemicals called neurotransmitters into the space outside the terminal where other neurons are located.

The transmitter diffuses in the space between the sending and receiving neurons and binds to receptors on the receiving neurons. The receptors are chemically linked to a specific transmitter—the transmitter functions like a key to open a lock. This arrangement thus adds an electrical step to the more typical way that all other kinds of cells communicate over short distances—by releasing chemicals. This offers a key advantage. Information can be transmitted across long distances in a short amount of time.

The space between the sending and receiving neuron is sometimes called a synapse. But this term more accurately refers to the connection between two neurons. Synapses have three components: the presynaptic site (the axon terminal of the sender that contains the neurotransmitter), the postsynaptic site (the location on the receiver that has receptors), and the tiny space between the presynaptic and postsynaptic elements, which is called the synaptic cleft.

The various neurons in the body collectively form the organism's nervous system. In the most fundamental sense, a nervous system is a sensory-motor integration device. Its job is to help an organism engage with its environment in the task of staying alive and well and reproducing its kind by identifying substances necessary for or detrimental to survival, and organizing appropriate responses to them. The inputs come in the form of messages received by sensory receptors for key classes of stimuli (light, sound, touch, odors, tastes), and the outputs involve motor effectors (muscles).

The most basic job of a nervous system is thus to connect sensory receptors with motor effectors. The simplest arrangement is one in which the receptors and detectors are directly connected; as we will soon see this is the case in the simple and diffuse nerve nets of Cnidaria. In complex animals, though, the nervous system is made up of billions of neurons with trillions of connections between them. But even in these, the nervous system exists first and foremost to help its possessor survive and thrive when engaging the environment by taking in sensory information and generating (and sometimes withholding) motor responses.

*Chapter 27*

# HOW NEURONS AND NERVOUS SYSTEMS HAPPENED

As with most evolutionary events, neurons did not suddenly appear. They were shaped in steps by small changes that took place during the transition from sponges to Cnidaria. Adult sponges, as we've seen, are sessile creatures that mostly stay attached to a fixed location. But in their youth, they are free-swimmers and get around quite a bit. The outer body surface of larval sponges is loaded with filaments called cilia, and used by the young organism to move around. These filaments are each attached to cells (one per so-called ciliated cell). They are similar to the flagellum of a choanoflagellate cell.

Two kinds of ciliated cells are found in different locations of the young sponge. Swimming cells have short cilia and cover most areas of the larval body. These beat constantly, causing random, undirected movements that keep the larva moving and afloat. Steering cells have long cilia, which are concentrated at one end of the body. They are sensitive to light, which causes the cilia to bend, directing movements toward the source of light. By the time the sponge becomes an adult, its cilia are gone.

Gaspar Jekely has proposed a fascinating hypothesis about how cilia-driven swimming in sponge larva paved the way for the emergence of neurons in Cnidaria. Jekely's idea is that neurons first arose to improve the efficiency of sensory-motor integration. Remember, choanoflagellates, which are closely related to the protist ancestor of sponges, have to detect light, swim, and steer (as well as feed and reproduce) with their single cell. Sponge larva have the advantage of having many cells, and their genes can divvy up chores among the cells. By separating light detection from general movement control, sponge larva separated sensory and swimming functions. But that left them with a problem. Sensory cells do not have a way to rapidly influence the swimming cells, since they are in different parts of the body, and chemical communication is too slow to do the trick. The solution was for the short-ciliated cells to keep the larva in constant random motion, and have the sensory cells with long cilia function as stimulus detectors as well as be in charge of steering, which is less demanding than swimming.

But clearly this proved to be an inefficient arrangement compared with what neurons ended up doing. So how did neurons and synapses develop? Jekely's hypothesis suggests a sequence of changes (figure 27.1). The first hypothesized transformation was the clustering of sensory and motor cells in the same vicinity, rather than in different parts of the body (as is the case for light detection/steering cells versus swimming cells). Chemicals released from sensory cells could then diffuse to adjacent cell bodies of motor cells

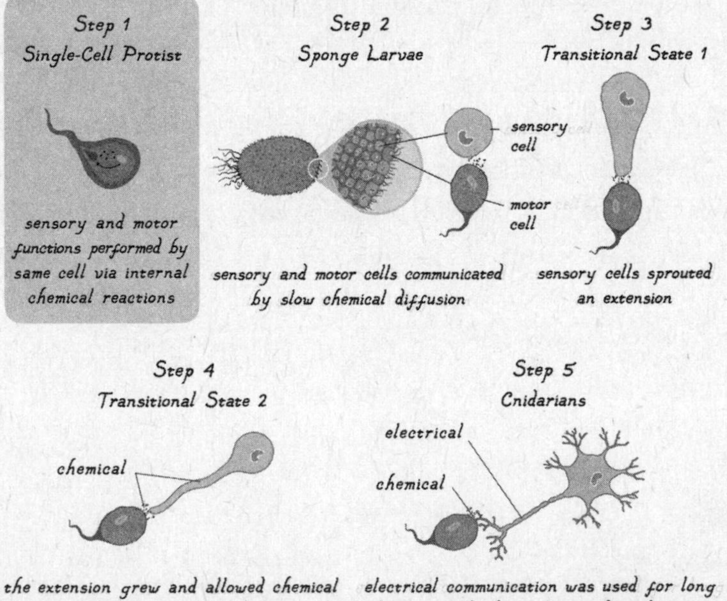


**Figure 27.1:** ***Steps from Chemical to Neural Communication***

and coordinate their activity, an adequate method for short distances. Next, by growing an extension of the sensory cell body outward, chemicals released by the sensory cell could affect motor cells a bit farther away. This helped overcome the spatial limits of chemical diffusion between cells a bit, but as the extension got longer, another limit was hit—communication still relied on slow chemical diffusion within the extensions of the sensory cell. The solution was the use of rapid electrical communication within the extensions, which became axons, and then, slow chemical communication to bridge the short space between the sensory and motor cell. As a result, the distance between sensory and motor cells became less of a factor, and cells in one part of the nervous system could communicate with cells in other body parts, regardless of how far apart they were.

It's natural to think of evolution in terms of its effects on adult bodies. But that makes it very hard to imagine how hydra and jellyfish could have resulted from sponges. That process begins to seem less inconceivable once we realize that sponges and Cnidaria both go through a ciliated larval stage and then both transition to a vaselike polyp shape (figure 27.2). In other words, genetic modifications of the sponge development plan could have, via natural section, produced larva that develop into novel polyps that were the foundation of Cnidaria.

In general, early life states tend to reflect the relation of the species to

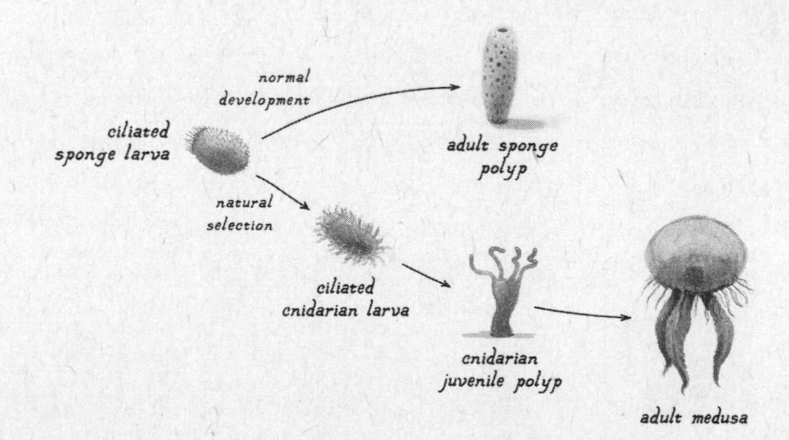


**Figure 27.2:** ***How Sponge Larva Became Jellyfish***

its evolutionary ancestors somewhat better than the form of the adult body. This is because natural selection changes the way organisms are built by genes during development. When developmental mutations are beneficial they become more frequent in the population, and when enough such individuals with this feature accrue, the Bauplan changes sufficiently that a new species or even phylum comes to be. The close connection of early development to evolution is part of what drives the field affectionately known as evo-devo (evolution and development).

There is another crucial part to this story of how sponges gave us nervous systems. Although they do not have neurons, they do possess what Seth Grant calls protosynaptic building blocks. Specifically, they have genes that, in later animals, are responsible for the presynaptic site (for example, genes for proteins that form structures that hold neurotransmitter chemicals in packets while they await release into the synaptic cleft) and for the postsynaptic site (for example, genes for receptors that bind released neurotransmitters), as well as genes for cell adhesion molecules used to stabilize synaptic connections once they are formed.

Why, despite having these essential ingredients, do sponges not make synapses? It appears that they lack the molecular signals that turn on the expression of the genes in a coordinated fashion during early development, in such a way as to generate a nervous system. In the absence of such a genetically encoded developmental program to direct the connection of presynaptic and postsynaptic elements, behavior could not be precisely controlled in response to sensory information. It is not sufficient for cells near each other to fuse and adhere for the animal brain to form. In order for synapses to be dedicated to vision, touch, or taste, and to be able to direct movements of specific parts of the body, or of the whole body, toward or away from a specific stimulus, precise wiring of connections between cells is required.

Particularly fascinating is the fact that some of these protosynaptic building blocks are also found in choanoflagellates. In these protists and sponges, the synaptic building blocks were not lying in wait to be used to make neurons and synapses, but were employed for other purposes. When the time came to make synapses between neurons to solve the problem of communication between different body parts, the existing building blocks were put to use for that purpose. And once the building blocks were co-opted by Cnidaria for synaptic transmission they were retained for this purpose in all animals thereafter.

The nervous system of Cnidaria is rudimentary, consisting mostly of a simple nerve net, a diffuse collection of neurons distributed through the outer skinlike tissue layer (figure 27.3). (Interesting factoid: the relation between skin and neurons persists in vertebrates like us, in which neurons and skin cells both arise from the ectoderm layer in the developing embryo.) Like all nervous systems that came later, cnidarian nerve nets are basically sensory-motor integration systems that do three basic chores. First, they receive messages from sensory receptors that detect light, touch, gravity, or chemicals. Second, they select and process the sensory messages. Third, they generate motor commands that control muscle movements. In this way, the various parts of the organism can respond as a unit. However, with a nerve net, responses are not easily localized. A hydra responds the same way regardless of what part of its body makes contact with a stimulus.

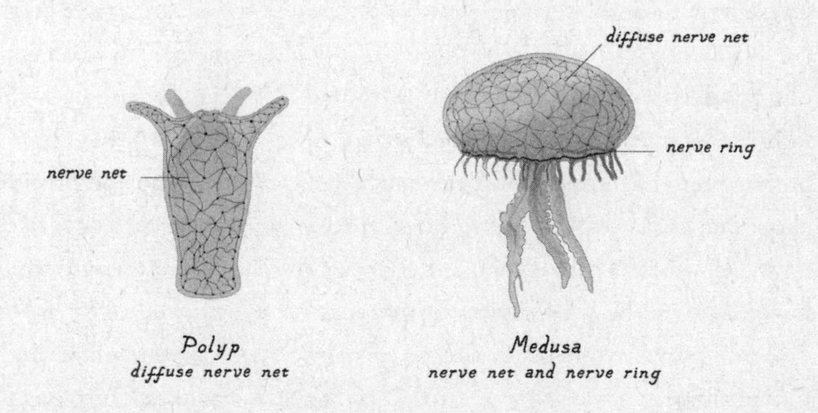


**Figure 27.3: *Cnidarian Nervous Systems***

While polyp-shaped Cnidaria like hydra mainly have a diffuse nerve net, jellyfish additionally have some concentrated collections of neurons. For example, some neurons form a neural ring around the medusa that controls its movements in fast swimming; others are clustered in the tentacles and are used to control slow swimming and prey capture, and in sperm transfer in sexual reproduction. These various localized neuronal collections thus allow more precise responses to stimulation than is possible with only a nerve net.

Detlev Arendt and colleagues suggest that genetic findings indicate the concentrated neurons in the umbrella and mouth of jellyfish are precursors of the more complex body and brain plans characteristic of the bilateral animals that evolved from Cnidaria. The umbrella concentration seems to be a forerunner to the specialized collection of neurons that appeared in the head region of most metazoans (in other words, a precursor to what we know as the brain). The other concentration, around the mouth and tentacles, appears to have elongated to become the nerve cord that connects the brain to the rest of the body (in vertebrates, this is the spinal cord).

The process I have described here suggests that the need for sensory and motor cells to communicate is the basis for how neurons came to be. We have so far defined the nervous system as basically a sensory-motor connection device, but can now expand that definition by saying that it is a collection of cells that sit between sensory and motor cells of the body, and coordinate muscle movements of the body in response to sensory inputs. In some organisms the neural intermediary is relatively simple (as in the nerve nets of Cnidaria), while in others it is extremely complex (the vertebrate brain, for example). The great neuroscience pioneer Sir Charles Scott Sherrington put it this way: "The brain seems a thoroughfare for nerve-action passing its way to the motor animal."

Yet if nervous systems existed simply to relay information from point to point, behavior would be limited to simple innate responses. One of the great advantages endowed to animals by the presence of a nervous system is the facility with which neurons can be modified when the organism interacts with its environment. This capacity, called synaptic plasticity, is the basis for learning.

It has been suggested that a key factor in the Cambrian Explosion of animal bodies was the advent of nervous system–based learning. While organisms without a nervous system can learn (recall the learning ability of single-cell microbes), nervous systems made learning much more sophisticated and flexible. And this revamping of the survival tool kit could have contributed to body plan diversification. For example, neural learning could have enhanced the ability to exploit new niches, and this would result in changes in body plan features needed to survive. Also, with both predator and prey capable of learning, the evolutionary arms race between them would have accelerated, resulting in unprecedented changes in body features. The importance of learning in survival has only increased with time and further Bauplan diversification.

## Part 6

# Metazoan Bread Crumbs in the Oceans

*Chapter 28*

# FACING FORWARD

Six hundred and fifty million years ago animal life was dominated by aquatic organisms with primitive (asymmetric or radial) bodies that either lacked a nervous system (sponges), or at most were in possession of simple nervous systems (cnidarians and ctenophorans). Then, around 630 million years ago, a new body style, bilateral symmetry, appeared. Bilaterals went on to acquire souped-up nervous systems that sported a collection of neurons in the head, a brain, which could evaluate the environment and behave in ways more complex than had ever been possible before. During the Cambrian Explosion, 543 million to 480 million years ago, brainy bilaterals living in the oceans rapidly became numerous and diverse in body plans. By 400 million years ago, some of these bilateral invertebrates, in particular millipedes, had invaded the land, and by 350 million years ago they had been joined by amphibians, the first vertebrates to live by breathing atmospheric oxygen. Evolution continued both under and above water, resulting in the great diversity of animals alive today. There must have been something special about a bilateral Bauplan that, once it arrived, caused it to become the main body plan that has dominated animal life ever since. The bread crumbs left behind by the early aquatic bilaterals thus offer important clues in our journey from LUCA to us.

Cnidarians, as discussed earlier, have a radial body with a central body axis extending from a single digestive opening (mouth) straight through the opposite end of the inner cavity. The opening is on top in polyps and on the bottom in medusae. In neither case is there a front-to-back axis, as they are more or less tubular (figure 28.1). This Bauplan worked well for their relatively low-mobility lifestyle. In stationary polyp bodies, the tentacles grab things that float by from any horizontal direction and hand them off to the upward-pointing mouth. Similarly, medusae can feed by drifting randomly with the current with their mouths facing downward, grabbing food from all directions. While medusae can slowly swim while "fishing" and can dart away from danger, for the most part they are fairly passive in their activities.

**Figure 28.1:** ***Radial Body Plan of Cnidarians***

Bilateral animals, by contrast, tend to be highly mobile, and have a preferred direction of locomotion. When they start moving, they go forward, rather than sideways or backward. But what is forward? Forward is a direction that emerges from the shape of a bilateral body.

You'll recall that bilaterals have two major axes: front to rear and top to bottom. These are more formally called anterior-posterior and dorsal-ventral axes (figure 28.2). In most animals, the long axis of the body, the anterior-posterior axis, is typically horizontal (parallel to the ground), with the anterior end defined by the head. The dorsal-ventral axis, on the other hand, is defined in relation to the belly and back; the belly is ventral (nearest the earth), while the back is dorsal (nearest the sky). Upright animals (humans) are the exception—as figure 28.2 shows, the axes are rearranged a bit and labeled somewhat differently.

Bilaterals were the first organisms to possess a head, and the direction

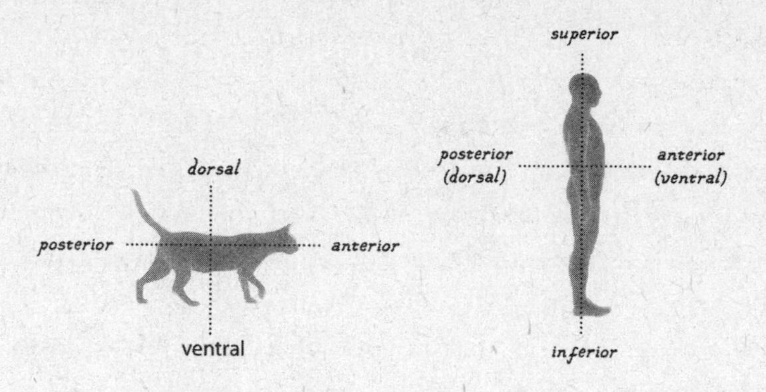


**Figure 28.2:** ***Body Axis Nomenclature in Horizontal vs. Upright Organisms***

that the bilateral head faces is forward, which is the most typical direction of locomotion in bilaterals. The head is also the location of many of the key sensory organs (eyes, ears, nose) that guide forward-moving behaviors in the search for, and approach toward, food, drink, shelter, or mates, and in detecting and moving away from danger. Predator-prey interactions especially ramped up in complexity in animals with bilateral bodies, allowing directed chase and escape.

The head of bilaterals also houses the brain, keeping sensory intake and processing and the coordination of behavioral responses of the body in close proximity in elongated bodies. With the mouth also located in the front of the head, the consumption of food and drink does not require a change of position. The brain is able to control the contractions of muscles in specific body parts because it is connected to a longitudinal nerve cord that runs through the anterior-posterior axis with nerves exiting all along the way. The brain and nerve cord of a primitive bilateral animal is shown in figure 28.3.

A popular theory holds that bilateral bodies emerged because of the advantages of a forward-facing body for locomotion in carrying out survival activities. While some have proposed other theories about why bilateral bodies first arose, most agree that the advantages conferred to locomotion, which is at the heart of many survival activities, account for

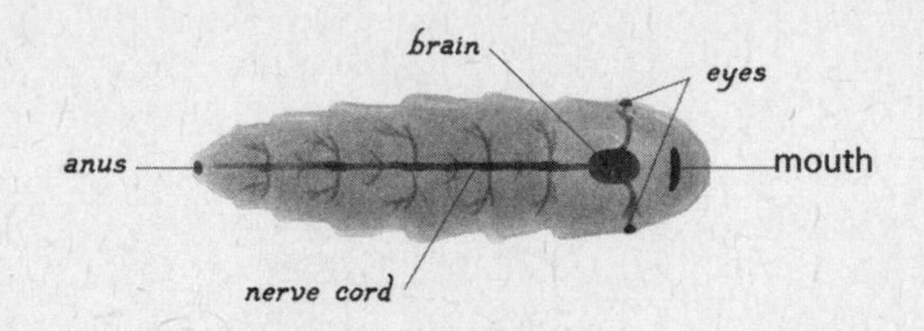


**Figure 28.3:** ***The Brain and Nerve Cord in a Primitive Bilateral Animal***

why the bilateral body plan has persisted as the ubiquitous default shape of animal bodies for so long.

Given that bilateral animals evolved from radial animals, a key question is how bodies with two axes emerged from those with a single one. Homeobox genes are major players in Bauplan construction in metazoans. Particularly important for bilateral animals is a subset of these called hox genes. Animals that lack bilateral symmetry have few of these. They come into their own in bilaterals, directing the construction of bodies with symmetry along the anterior-posterior axis and guiding the development of body parts along this axis, a process called anterior-posterior axial patterning. We will have more to say about hox genes when we discuss the evolution of vertebrates.

*Chapter 29*

# TISSUE ISSUES

Eumetazoans, you'll recall, are animals that are defined by the presence of tissues—all radial and bilateral animals are included. Tissues result from a process of cell specialization. After each of the daughter cells of a fertilized egg divides, it gives rise to identical cells that collect together in a single layer that forms a hollow sphere called the blastula. The blastula then folds inward, forming a structure called the gastrula, with distinct layers of embryonic cells. From these are made all the specialized cells of the various tissues and organs of an animal's body, except for gametes (recall the difference between somatic and germline cells).

Radials are diploblasts, as they have two embryonic cell layers, while bilaterals are triploblasts, having three (figure 29.1). The two embryonic cell layers in radial eumetazoans, such as cnidarians, are called the ecto-

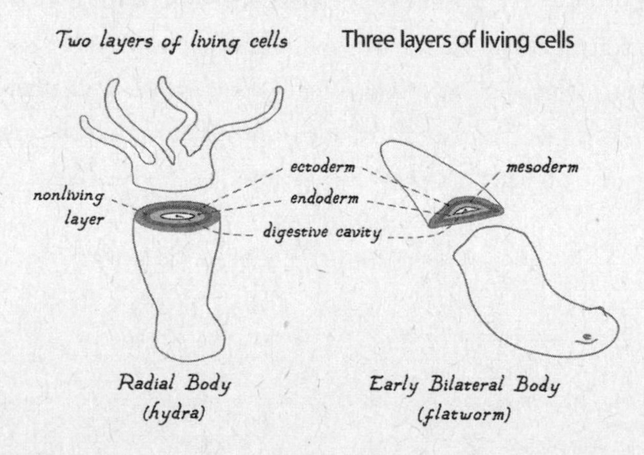


**Figure 29.1: *Tissue Layers in Radial and Bilateral Animal Bodies***

derm and endoderm. They give rise to the two baglike compartments of the body—the outside covering or skin and the gut (like the cell layers, these body layers are also called the ectoderm and endoderm). Radials have a third body compartment, the mesogel, but it is not made of living tissue.

Like radials, bilaterals have an ectoderm embryonic layer responsible for cells in the outer layer (the skin) and an endoderm layer that makes cells that form the gut. But they also have a third embryonic layer, the mesoderm, which lies between the ectoderm and endoderm and generates additional cells. It is responsible for a tissue structure present in most bilaterals, but that radials lack—the coelom, a body cavity in which the gut and other inner organs reside.*

Several specific tissues are especially important in distinguishing bilateral from radial animals. Radial guts have one opening, which serves as a mouth and anus, but the bilateral eumetazoans that evolved from radial cnidarians separated input and output, having guts with the mouth at one end and an anus at the other. Second, as noted, only bilaterals have a coelom to house internal organs. The third tissue distinguishing radial and bilateral animals is the nervous system. In cnidarians, as we've seen, the nervous system is relatively primitive and either lacks, or has minimal, centralized control over body movements—there is some motor coordination but not by way of orders from headquarters. Most bilateral animals, in contrast, have centralized control in the form of brains that integrate information from sensory organs, such as paired eyes in the head, to guide movement of specific body parts in the process of initiating and performing complex movements that engage specific body parts or the whole body.

* Each of the three embryonic cell layers in bilaterals has to generate many additional kinds of cells and tissues. For example, in vertebrates, the ectoderm accounts not only for skin, but also skull and teeth. A special part of the ectoderm called the neuroectoderm generates neurons that form the nervous system. The endoderm generates not only the gut wall but also some other inner organs, such as the liver and lungs. The mesoderm cells, besides making the coelom in bilaterals, also generate the skeletal bone cells, cartilage, muscles, gonads, kidneys, heart, blood, and lymphatic system.

TABLE 29.1: Defining Features of Typical Bilateral Animals

| Defining Features |
|---|
| Two body axes (front to back and top to bottom) |
| Symmetric halves along the front-to-back axis |
| Three body layers (ectoderm, mesoderm, endoderm) |
| Gut with oral and anal openings |
| A cavity (coelom) between the gut and body wall |
| Localized sense organs, many in the head |
| A nervous system with a central control unit (brain) |

In attempting to obtain clues about the organism that was the last common ancestor of all bilaterals (LCBA), scientists have searched for fossils and/or molecular evidence that account for features of bilaterals. Both approaches suggest that the LCBA emerged about 630 million years ago. A leading hypothesis based on molecular evidence is that the LCBA was a primitive aquatic worm with a bilateral body plan. While subsequent bilaterals possessed sensory organs and a brain at the anterior end, a gut with an anterior mouth opening, and a posterior anus and a coelom, fossil evidence suggests that the LCBA lacked some of these features. Its nervous system may have been relatively diffuse, its gut not fully formed, and it lacked a coelom. But these features were quickly acquired by its descendants.*

A reconstruction of what this animal might have looked like is shown in figure 29.2. In considering how this kind of creature might have come to exist, scientists again turned to larva. Free-swimming larva of cnidarians exhibit some degree of bilateral symmetry, and have many of the genes that are used to direct body development in bilaterals.† This suggests that the LCBA emerged as a modification of the cnidarian larval body, and

* Some living bilaterals do not have a true coelom or brain; their ancestors are believed to have lost these features.
† Because of the bilateral features of cnidarian larva, some have proposed that Cnidaria as a group should be considered bilaterals. However, this idea has not gained full support. Since our path toward human nature does not change significantly by whether or not wormlike creatures were the first bilaterals, I stick with the more conventional view.

inherited a diffuse nervous system and an incomplete gut—it was not quite a full-blown bilateral, but sufficiently different from a cnidarian to count as a bilateral. Just as sponge larva made Cnidaria possible, cnidarian larva gave us the bilateral branch of the tree of life, and thus most of the animals we live with today.

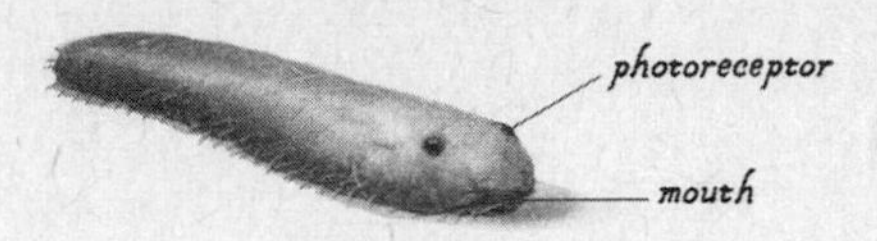


**Figure 29.2:** ***Reconstruction of the Last Common Bilateral Ancestor***

*Chapter 30*

# ORAL OR ANAL?

The most common way to group bilateral animals is by whether or not they possess a vertebral column—a spine. Vertebrates have one, and invertebrates don't. While this distinction is an important part of our story, there is a more fundamental one that helps explain the evolutionary path from the last common bilateral ancestor to vertebrates.

Recall that in jellyfish and other radial metazoans the digestive tract has one opening that serves as both an input and output channel. By contrast, in typical bilateral animals the mouth and anus occupy different ends of the digestive tract. And the way these two openings progress during embryonic development is the key to the most fundamental division of bilaterally symmetric animals.

During the early life of a bilateral animal, as the blastula folds inward to form the gastrula, the fold gives rise to an opening that becomes the digestive tract (figure 30.1). In some, this first opening of the blastula becomes the mouth end of the digestive tract. Such organisms are called protostomes (from Greek, meaning "mouth first").

The vast majority of invertebrates belong to this mouth-first group: insects (flies, bees), arachnids (spiders), crustaceans (crabs, lobsters), mollusks (snails, slugs, clams, squid, octopi), and a variety of worms. In the remaining bilaterally symmetric animals the first opening becomes the anus. Anus-first animals are known as deuterostomes (literally, "mouth second"). Among these anus-first organisms are a unique group of invertebrates that were the ancestors of vertebrates (figure 30.2). This embryological difference between protostome and deuterostome invertebrates is

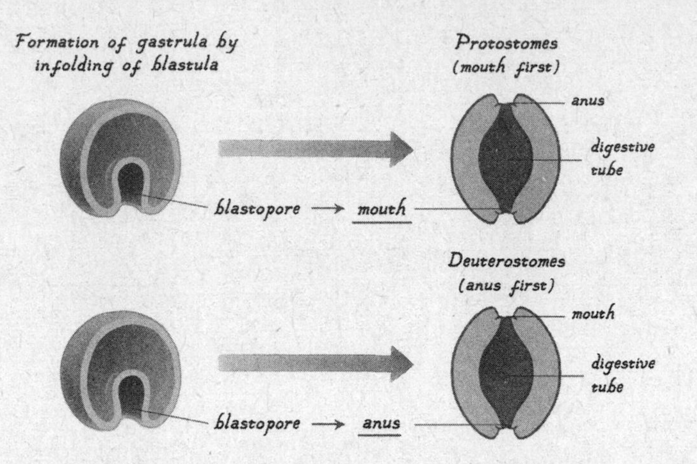


**Figure 30.1:** ***Digestive Tube Formation in Protostomes and Deuterostomes***

what enables us to connect vertebrates to their invertebrate ancestor, and thus to the deep history of life.

By 580 million years ago both protostomes and deuterostomes had appeared. It was long thought that their common ancestor (the protostome-deuterostome ancestor, PDA) was the same as the LCBA. As we saw in the last chapter, the LCBA, which broke away from the radial body plan some 630 million years ago, had only some features that are present in most later bilaterals. The PDA thus came after the LCBA but before the split between protostomes and deuterostomes. The LCBA and PDA are both considered early or basal bilaterals (see figure 30.2). The exact nature of the PDA is not known. But it was likely a member of a group called Nephrozoa, the clade in which a coelom first appears, and includes all protostome and deuterostome animals, and thus practically all existing animals.

While we share many more genes with other deuterostomes, we have a good number in common with protostomes as well, since both groups received key survival-related genes from the PDA. This accounts for why research on fruit flies, worms, sea slugs, and other protostomes can reveal important information relevant to humans, and is being used to under-

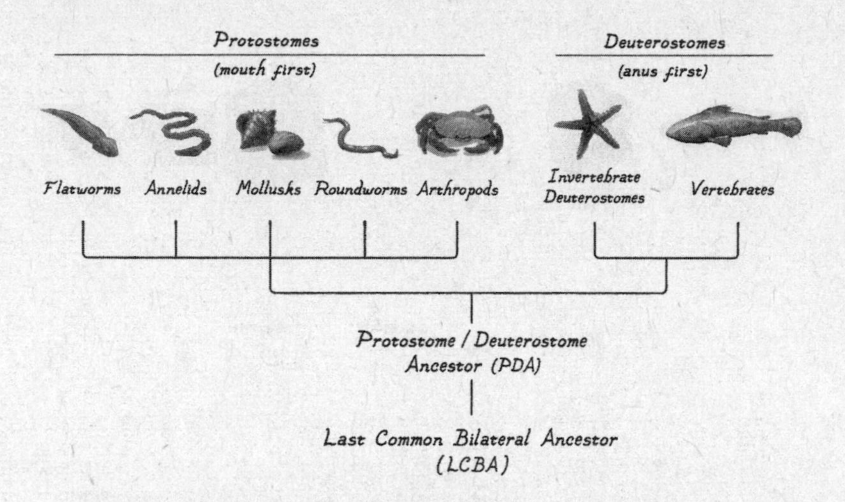


**Figure 30.2:** ***Common Origin of Protostomes and Deuterostomes***

stand and develop treatments for such human diseases as cancer, heart disease, diabetes, and neurological disorders, including memory disorders. For example, Eric Kandel's groundbreaking and Nobel Prize–winning discoveries about the molecular mechanisms of long-term memory storage in sea slugs have been shown to apply to mammals, including humans.

Protostomes are divided into two groups, one of which includes flatworms, annelids (segmented worms) and mollusks (clams, oysters), and the other arthropods (insects, arachnids, crustaceans) and roundworms. These are defined by whether the animal's body grows continuously or whether it sheds its body (molts) and starts fresh at some point. Genetic analysis shows that the annelids and mollusks share more genes with deuterosomes than do the arthropods and roundworms. This suggests that annelids and mollusks offer advantages as model systems for understanding human function and disease, in spite of the fact that insects (fruit flies) and roundworms (*Caenorhabditis elegans*) have, for practical reasons, been especially popular as research subjects in genetic studies.

*Chapter 31*

# DEEP-SEA DEUTEROSTOMES LINK US TO OUR PAST

I remember how excited my young sons once were to come across a starfish on the beach. Immobile, and partially buried in sand, the creature seemed part and parcel of the shells and other sea debris washed up on the shoreline. The boys were shocked, then thrilled, when, while holding the treasure, its tentacles started moving.

With tentacles extending in various directions from a more or less circular (radial) body, starfish appear to be a flattened-out version of a medusa or hydra. But they are actually deuterostomes. This may be surprising, given that deuterostomes are bilaterally symmetric animals, and adult starfish do not have that characteristic. However, in early life, as larvae, they are indeed bilaterally symmetric, a feature they obviously lose as they mature (figure 31.1). As we've seen, the larval state is often more directly related to the organism's evolutionary history than the adult state.

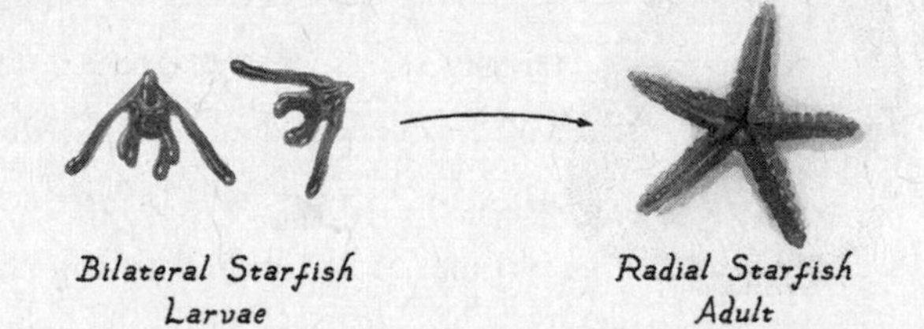


**Figure 31.1:** ***Starfish Begin Life with Bilateral Bodies***

There are five deuterostome phyla (figure 31.2), which fall into two groups, or superphyla. One is Ambulacraria, which includes members of phylum Echinodermata, such as starfish, as well as members of phylum

Hemichordata, which includes acorn worms. The remaining three groups of deuterostomes are in a second superphylum, Chordata.* Two of the Chordate phyla contain invertebrate deuterostomes. These are Urochordata, commonly known as sea squirts, and Cephalochordata, known as lancelets. The third chordate phylum is Vertebrata, home to fish, amphibians, reptiles, birds, and mammals—including humans.

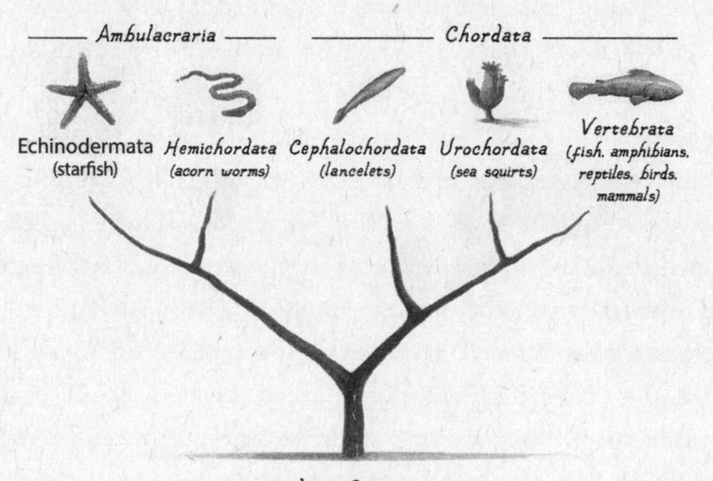


**Figure 31.2: *Two Deuterostome Superphyla***

It was once thought that ambulacrarians were the first deuterostomes and that chordates evolved from them. Given that vertebrates are chordates, this scenario would suggest that ancient starfish might have been part of the ancestry of humans and other vertebrates. More recent evidence, however, suggests that Ambulacraria and Chordata are sister groups that evolved through separate lines from the last common deuterostome ancestor (LCDA) during the Cambrian Explosion. Although starfish are not in our direct lineage, as deuterostomes, they are still close relatives.

In some sense we've answered the question about the origin of invertebrate chordates (which evolved from basal deuterostomes that emerged as

* Until recently Chordata was considered a phylum and the three groups within it were subphyla. I adopt the newer convention here, treating Chordata as a superphylum made up of three phyla.

a result of the protostome-deuterostome split) and also of vertebrates (which evolved from invertebrate chordates). But there are two invertebrate chordate groups—which raises the question of how the first vertebrates relate to each of them.

The short answer is that urochordates, also known as tunicates, are a sister group to vertebrates, and cephalochordates are a sister group to both of them. Because urochordates are thus somewhat closer to vertebrates than cephalochordates, it might suggest that we can learn more about the evolution of vertebrates from present-day sea squirts than lancelets. But as the evolutionary neurobiologist Linda Holland points out, the basic genetic tool kits of vertebrates and cephalochordates evolved slowly after they diverged more than 500 million years ago, while the urochordate genome has undergone more extensive transformations. For this reason present-day cephalochordates better represent the ancient chordate history and are thus a better model of vertebrate origins than urochordates.

In the next chapter we will thus focus on lancelets—small, wormlike cephalochordates. Though they are active free-swimmers, moving with undulatory-type movements, they mostly live at the bottom of the ocean, and swim mainly to locate places where they can live a stationary lifestyle, and/or to escape from predators. They are themselves predators, but do not hunt and feed while swimming. Being filter feeders, they acquire food while buried in the sand and gravel of the ocean floor; with their mouth sticking up, they suck water in to extract nutrients. What makes these invertebrate chordates so important to vertebrate evolution is a unique structure in their backs, a structure from which vertebrates evolved their defining vertebrate column.

*Chapter 32*

# A TALE OF TWO CHORDS

One particular feature of chordates—and one that no other kind of organism has—is a notochord, which gives this group their name. *Noto* means "back," and the notochord itself is a flexible, hollow rod made of cartilage that runs through the long axis of the chordate body along its dorsal side, or back. The notochord functions as a primitive skeleton that supports the body against the pull of gravity—the skin and some inner body parts hang from it the way clothes hang from a rod in a closet. The gelatinous filling at its hollow center enables its body to be flexible and to engage in undulatory swimming movements.

Like many features in evolution, the notochord may have come about as a modification of a previously existing structure. Genetic evidence suggested that the changes were to a muscular sheath running through the long axis of the LCBA or primitive worms closely related to it.

Protostomes and deuterostomes, as we've seen, both descended from the PDA, which is itself a descendant of the LCBA. Present-day chordates and protostomes have an interesting connection that has been illuminated by studies of annelid worms, ancestors of which were early protostomes. They possess a longitudinal muscle, called the axochord, which has genetic and morphological characteristics that overlap with the notochord of chordates. From this similarity, biologists have argued that the notochord of chordates and the axochord of early protostomes have a common origin, the long muscular sheath of the LCBA, which was discussed earlier.

Another significant point is that the LCBA, or an early descendant of it, is believed to have had a longitudinal nerve cord with a specialization

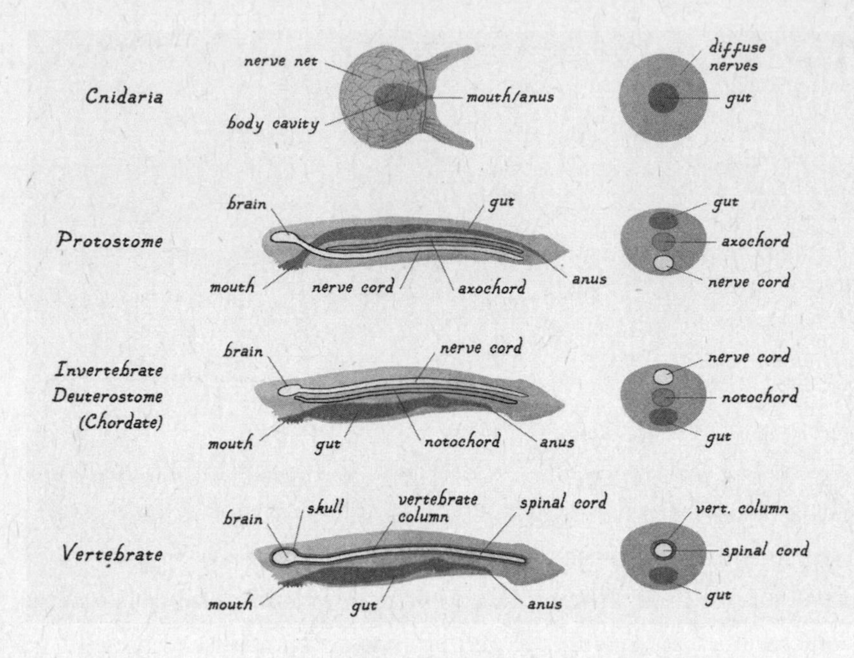


**Figure 32.1:** ***Two Cords/Chords in Protostomes and Deuterostomes***

(brain) at the anterior or head end. It is thus conceivable that chordates and protostomes both inherited their nerve cord and brain from these early bilaterals. But there are some facts that are potentially problematic for this conclusion.

For one thing, the chordate nerve cord is dorsally situated above the notochord, with the gut below both (figure 32.1). In protostomes, however, the nerve cord lies below the axocord, and both are below the gut. These two nerve cords may have had nothing to do with each other, but Linda Holland believes they did.* She cites extensive and compelling evidence involving hox and other genes showing deep connections between the protostome ventral nerve cord and the chordate dorsal nerve cord, and concludes that the nerve cord was located ventrally in the common ancestor of protostomes and deuterostomes, the PDA, and stayed down south

* I am grateful to Linda Holland for an email exchange on this topic.

when protostomes diverged from the PDA, but moved north with the evolution of chordates. This is not a far-fetched notion, as it is natural for things to move around as body plans adapt via natural selection.

Another potential problem with the idea that chordates inherited their central nervous system from the PDA involves the nature of the nervous systems possessed by invertebrate deuterostomes. The cephalochordate lancelet, for example, has a dorsal nerve cord, but not much of a brain. There is some swelling at the anterior end of their nerve cord, but, given its minimal size, lancelets have often been described as brainless. Nevertheless, recent evidence from studies comparing specific cellular structures in the lancelet larva with those of vertebrates has provided morphological evidence for the existence of neural equivalents of a small but bona fide brain in lancelets that has parts that are homologous with lower parts of the vertebrate brain. This conclusion is strongly backed up by findings showing that hox and other genes present in neural structures in vertebrates are also present in the equivalents in lancelets. Some of these genes also have counterparts in protostome brains, suggesting that these neural genes were inherited by both groups from the PDA, which likely got them from the LCBA/PDA. While urochordates are more closely related to vertebrates than cephalochordates, as described in the previous chapter, urochordates have undergone much more diversification over time than cephalochordates, making the latter a useful model of early chordate origin of vertebrates.

Putting all this together, we can conclude that protostomes and cephalochordate deuterostomes inherited from their common ancestor (the PDA) a central nervous system consisting of a brain and nerve cord, as well as a structural support chord (an axochord in protostomes, and a notochord in chordates). With the evolution of vertebrates, the notochord gave way to the spine (vertebral column), and the dorsal nerve cord became the spinal cord. In vertebrates, the small anterior expansion of the nerve cord seen in lancelets retained key structural features controlled by conserved genes in its deuterostome ancestors. In spite of these impressive correspondences with the brains of other chordates and even protostomes, the vertebrate brain, as we will explore in the remainder of this book, is an unparalleled development in the history of life.

You may be wondering, given that the notochord is a key feature defining the chordates, and that the notochord was ultimately replaced by the spine in vertebrates, why vertebrates are still considered to be chordates, rather than constituting a completely separate group. In embryonic life, all vertebrates possess a dorsal notochord, which is then replaced by a vertebral column as the embryo matures. The gelatinous interior of the notochord becomes the soft material inside discs between vertebrae. When a disc is herniated, the material squeezes out, and with the loss of this cushion, inflammation and compression of nerves can result, leading to sciatica and back pain. Also, all chordates, including humans, possess pharyngeal arches—swellings and indentations in the neck region. In ocean-dwelling organisms, including invertebrate chordates (lancelets) and underwater vertebrates (fish), the indentations open and the arches become the gill apparatus. In land vertebrates, which get their oxygen from the air through lungs, the arches are repurposed during early life to become jaws and parts of the inner ear. We are now ready to explore the vertebrates.

Part 7

# The Vertebrates Arrive

*Chapter 33*

# BAUPLAN VERTEBRATA

Depending on how you count, there are roughly twenty-eight bilateral phyla with distinct Bauplan features. Twenty-seven of these are invertebrate phyla, including twenty-three groups of invertebrate protostomes and four invertebrate deuterostomes. There is only one vertebrate phylum;* the vertebrate Bauplan stands alone, singular among many invertebrate ones (table 33.1).

TABLE 33.1: Defining Bauplan Features of Vertebrates

| **ADULT FEATURES** |
| --- |
| —Vertebral column |
| —Endoskeleton anchored on vertebral column |
| —Skull with braincase anchored to anterior end of vertebral column |
| —Well-developed brain |
| **EARLY LIFE FEATURES REFLECTING CHORDATE HERITAGE** |
| —Hollow nerve cord under dorsal surface of the body |
| —Notochord under nerve cord and dorsal to gut |
| —Pharyngeal gill slits |
| —Postanal tail |

Though the vertebrate body is unique, it evolved from the Bauplan of invertebrate chordates. And just as chordates get their name from their

* Recall from chapter 31 that here we adopt the newer convention of treating Chordata as a superphylum and Vertebrata as a phylum (as opposed to Chordata being a phylum and Vertebrata a subphylum).

notochord, vertebrates take theirs from their vertebrae, the small bones that form the vertebral column (aka spine, or backbone), the structure around which the vertebrate body is built.

The flexible cartilaginous notochord enabled lancelets to swim with slow, undulatory-type movements, useful for getting around but not ideal for capturing prey while swimming. Lancelets are predators, but as noted do not hunt while swimming; instead they feed by sucking in water while mostly buried in the ocean floor. To hunt the ocean waters requires the capacity for fast, agile swimming, both to catch prey and avoid being eaten. Having a body built around a structure with individual parts (vertebrae), each of which was movable, provided the flexibility that typifies vertebrate behavior.

The vertebral column surrounds the spinal cord, which is, of course, a version of the nerve cord that runs lengthwise through the body, from neck to tail, of invertebrate chordates. The neck end of the spinal cord is continuous with the brain, which is enclosed in a braincase, a part of the skull, which is itself an extension of the vertebral column. The vertebral column is an anchor to which the rest of the skeleton (including the tail, rib cage, and limbs) is attached. It also anchors the muscles and internal organs, just as the notochord did for invertebrate chordates. Many of these features are observable when you filet a fish and remove the innards from within the rib cage, and then, after cooking the fish, you separate the bones from the meat by lifting the tail and pulling the spine off the flesh, and continuing to also remove the head.

All vertebrates share key features listed in table 33.1. But each of the five classes is distinct. Indeed, vertebrates are an incredibly diverse group. They began aquatically, but some then evolved Bauplans that enabled them to live and move around on the land, and even in the air. The diversity of vertebrate Bauplans is evident from a perusal of the variety of skeletons they possess (figure 33.1). Body diversity and behavioral flexibility allowed vertebrates to reside in a wide range of climates across all regions of the planet.

An animal's Bauplan unfolds in early life under the control of its genes.

**Figure 33.1:** ***Vertebrate Skeletons Reveal the Diversity of Their Bauplan***

As noted earlier, a family of genes called homeobox genes are major players in body construction. These are ancient, having contributed to body design in all multicellular organisms, including plants, fungi, and animals. The subset of these called hox genes, discussed above, play key roles in building bilateral bodies.

Hox genes direct the construction of basic structural features that all protostomes and deuterostomes, all bilaterals, share, such as symmetry along the anterior-posterior axis. They get this job done by regulating other genes that initiate the construction of specific structures, such as legs and arms, up and down the body axis at specific times in embryonic life. Hox genes are responsible for internal organs as well, and also control the size of different body parts. As we saw in the last chapter, they are also crucially involved in building nervous systems.

Conservation of hox genes across phyla is what gives all bilaterals their common Bauplan features. Identifying similarities between these different groups of organisms helps uncover evolutionary relationships. For example, hox gene similarities helped scientists connect vertebrates to lancelets, and then lancelets, and hence vertebrates, to their deuterostome ancestors, and also to protostomes, and both to the LCBA.

It is the variations in the expression of genes between different phyla that contribute to their unique body designs. Of particular note is the fact that protostomes and invertebrate deuterostomes have only one set (cluster) of hox genes, while vertebrates have four. In other words, with the evolution of vertebrates from invertebrate chordates, hox genes were duplicated several times. Gene duplications of this sort are an important way that vertebrates have added Bauplan complexity.*

* While the complexity of vertebrate bodies relative to invertebrate ones is clearly related to gene duplications, there is not a simple one-to-one correlation between hox gene complexity and body complexity. For example, some less complex early vertebrates (fish) have slightly more hox genes than more complex later ones (mammals).

*Chapter 34*

# THE LIFE AQUATIC

Vertebrate life emerged in the oceans. While fish are usually said to have been the first vertebrates, and they are indeed the oldest living vertebrates, they were preceded by a creature that was intermediate between invertebrate chordates and fish. A fossil dating from 530 million years ago identified this creature, which was called *Haikouella,* after the location in China where it was found. The fossil revealed a structure that was transitional between a notochord and a primitive vertebral column—a notochord with segments. *Haikouella* doesn't have all the characteristics of vertebrates shown in table

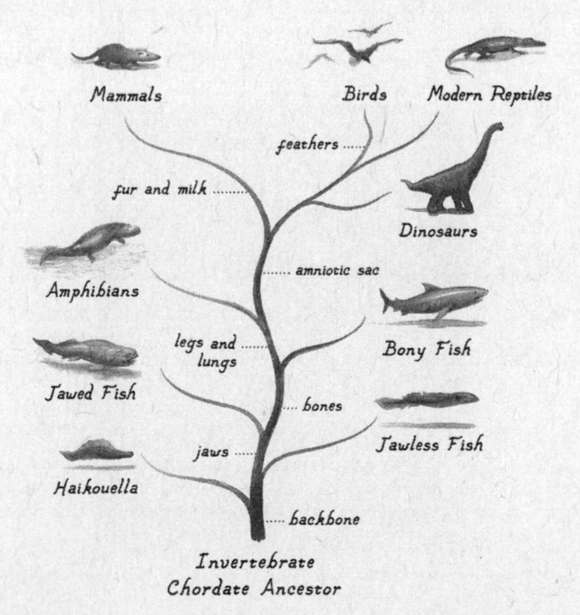


**Figure 34.1:** ***Vertebrate Family Tree***

33.1, but it goes far enough beyond the basic chordate features to count as a vertebrate. The vertebrate timeline thus starts with *Haikouella* (figure 34.1).

Fish appeared roughly 520 million years ago during the Cambrian Explosion. The first fish did not have bones, but skeletons made of cartilage, so this subgroup is called cartilaginous fish.

Cartilaginous fish initially lacked jaws. Their mouth was always open, with stationary primitive teeth that functioned as a filter. Unlike the passive filter-feeding behavior of lancelets, these jawless creatures were active predators. They ate while swimming forward, sucking water filled with ocean detritus through their teeth, or while diving down to the sediment and pulling in sand filled with organic material, or by attaching their mouth to larger prey and sucking out nourishment. They breathed through gills, which, as in earlier chordates, extracted oxygen from the water. Lampreys and hagfish, key examples of this group, have changed little over their millions of years of existence.*

Around 510 million years ago an important modification of the early vertebrate Bauplan occurred when parts of the gills of cartilaginous fish were converted into jaws fitted with powerful muscles and teeth (figure 34.2). Some jawed cartilaginous fish, like rays and skates, had flat teeth, enabling them to crush prey, both killing and mashing it before digesting it. Others, like sharks, had many sharp teeth attached to their jaws, making them efficient, ferocious predators, a reputation that persists today.†

By far the largest category of fish is the subset of jawed fish whose skeletons are made up of bones. These bony fish appeared around 480 million years ago. Unlike cartilaginous fish, which are all carnivorous predators,

---

* In November 2017, I visited Sten Grillner at the Karolinska Institute in Stockholm. He has spent much of his distinguished career studying lamprey. I asked to see one, and Sten readily obliged. The tank seemed empty, but when his colleague moved a net through the sandy bottom, the creature shot out and darted around the tank with rapid undulations. It was an exhilarating experience to see this ancient king of the sea in action, a fitting conclusion to my research on the early origins of vertebrates.

† An early extinct group of jawed fish are the Placoderms, which sported bony armored plates on their necks and heads. These plates are thought to have evolved as protection, since this was a time of rapid animal diversification, including fast and efficient predators, like sharks, with which they shared the oceans. And like the sharks, their skeletons were cartilaginous. In spite of their armored head, their body was largely unprotected, except with scales in some cases, and they were no match for the other predators of their day—sharks still exist but Placoderms are a thing of the past.

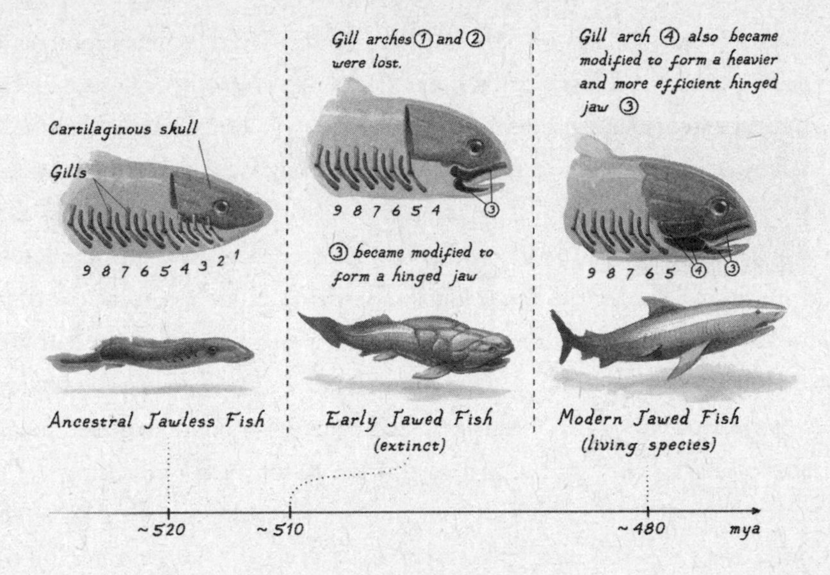


**Figure 34.2:** ***Jaws Evolved from Gills***

bony fish have more varied diets, with some being carnivores, and others herbivores or omnivores. Early bony fish diverged into two groups, one of which includes most of the animals commonly referred to as fish today (perch, bass, salmon, mackerel, piranha, and barracuda, to name some) but are technically called ray-finned fish. Their fins are fanlike or raylike, and composed of numerous small bones covered with a thin layer of skin.

Fish breathe by extracting oxygen from water using their gills. Gills, as we've seen, are a universal feature of vertebrate embryos. Bony fish not only use gills for breathing, but also for filling their swim bladder with air, giving them buoyancy and enabling them to float when not moving. While bony fish can breathe while stationary, cartilaginous fish, such as sharks, must keep moving in order to pass oxygen-filled water over their gills, and lacking a swim bladder, they sink when not swimming. Because of their enforced hyperactivity, it is occasionally said that sharks never sleep. But some sharks take a break by remaining still on the ocean floor, facing into a moving current of water.

The other group of bony fish is called lobe-finned fish. These split from the ray fins about 440 million years ago and are mostly extinct, but are of great interest because they are the kind of fish from which subsequent vertebrates evolved. A key characteristic is their possession of fins made up of a single bone surrounded by hunks of muscle tissue, giving the fins a "lobular" look (see figure in next chapter). They had two paired sets of these lobes on the ventral (bottom) side of their bodies (one pair in back and the other in front), and could use them to walk, stiltlike, across the ocean floor, and plant their position in currents. These structural modifications aided their ability to hunt aquatic invertebrates (mussels, shrimp, squid) or even other fish. The lobes of lobe-finned fish were the precursors of the muscular, jointed limbs (legs and arms) that all subsequent vertebrates possessed, a feature that helped make possible life on land.

*Chapter 35*

# ON THE SURFACE

About 375 million years ago, something important for us happened to fish: they began to diverge to form a new class of vertebrates that could live on land. The first step in this process, as described in Neil Shubin's book *Your Inner Fish*, was a "fishapod" (figure 35.1). This extinct creature was a transition between fish and tetrapods, the four-legged animals that would follow: amphibians, reptiles, birds, and mammals.

**Figure 35.1: *Fishapod***

The fishapod was essentially a lobe-finned fish equipped with some novel Bauplan accoutrements. Joints were added to the stiltlike limbs of the lobed ancestors, enabling smooth walking movements on the bottom of the body of water in which they lived. Also, in addition to gills, they possessed primitive lungs, which enabled them to breathe air from the atmosphere and occupy shallow, warm waters with low oxygen content. It was, in a sense, a small step to take for the descendants of fishapods, amphibians, to be able to expand their territory to the surface of the Earth, and therefore become the first terrestrial tetrapods.

As their name implies, amphibians were amphibious—able to inhabit land or water. Frogs, toads, salamanders, and the like are familiar examples. They start life as tadpoles that live underwater as vegetarians, using gills to extract oxygen, and develop lungs as they transition to their adult, carnivorous form. But amphibians could not have left the sea to live on land without the help of another multicellular group that turned terrestrial earlier and released sufficient oxygen into the atmosphere.

Plants had been expelling oxygen on land as a by-product of photosynthesis for about 50 million years before the amphibians walked ashore. This made it possible for respiration, and hence metabolism, to occur on land. But plants were also important for another reason, as they made herbivorous animal life possible. While amphibians didn't eat vegetation, some of their prey did.

The first animals to invade land were not carnivorous vertebrate tetrapods, but invertebrate protostomes, and specifically millipedes, which were vegetarian and lived off early plants—mossy growths on rocks. Millipedes belong to phylum Arthropoda, which share a common ancestor with ancient aquatic annelid worms and mollusks. In response to predation from protostomes, plants were pressured to diversify, which was made possible by the increased carbon dioxide in the atmosphere that resulted from animal respiration. With more types of plants to eat, terrestrial protostomes also diversified, in the form of insects and spiders. Since there were no other animals on land, amphibians had no competition for protostome prey, placing them at the top of the terrestrial food chain. As vertebrate life later diversified on land, carnivorous dietary opportunities opened up, but so did opportunities to be eaten.

Although amphibians can live in the water and on land, they must remain close to water in lagoons or ponds. They, like fish, reproduce sexually, but without sexual intercourse. Fish and amphibian eggs are encased in jellylike membranes and are laid in the water (though some amphibians also lay eggs in moist soil). Males then release their sperm upon them, a process called external sexual reproduction. The fetus develops in the soft jelly shell, and hatches into the world.

By 330 million years ago a new class of tetrapods had emerged. These were amniotes, animals in which the fetus develops in an internal, fluid-filled amniotic sac and with lungs that extracted oxygen from the air rather than water. This allowed vertebrates—in fact, animals—for the first time to give birth away from water, since the fetus matured inside the female. During pregnancy, the amniotic sac turned into a hard shell that was then "laid" on land, and the fetus then continued to grow in this egg until hatching. But this required a means of fertilizing eggs inside the female. In other words, it required external genitalia, the tissues that make possible sexual intercourse. Genitalia-bearing amniotes include all the remaining vertebrates—reptiles, birds, and mammals.

First came reptiles. Their long limbs made them more mobile on land than amphibians. They eventually split into two groups, one of which evolved to give rise to birds, and the other to mammals. In both birds and mammals the spatial range of foraging greatly increased, putting additional pressure on the brain to expand to accommodate the challenges that involved.

Synapsids were the first group of reptiles to branch off from the basal amniotes, about 310 million years ago (figure 35.2). These are known as mammal-like reptiles, and indeed were the ancestors of mammals. They included fierce predators that dominated the Earth for millions of years, as well as herbivores. Some were quite large, with fins on their back and bony armor; others had saber teeth. A second group split off from the amniotes a few million years later. These early sauropsids tended to be small, and were no match for the synapsids of the day. Then, around 250 million years ago, the Earth heated, and much of the animal and plant life on land and in the seas perished in this mass extinction.

For those species that survived extinction, new opportunities arose. As Michael Benton notes, that was the case for some synapsids and sauropsids. The most important synapsids that endured were the cynodonts.

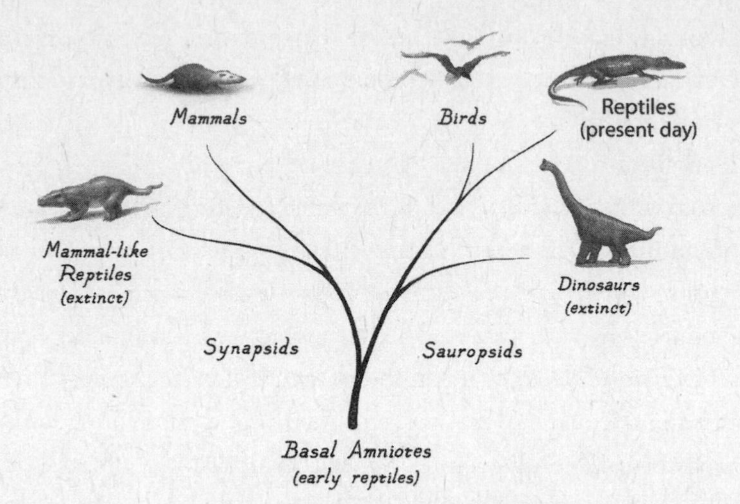


**Figure 35.2:** ***Reptilian Family Tree***

They were about the size of a large dog, and had features that, in retrospect, showed more and more hints of the mammals to which they would give rise, including changes in the jaws, eye sockets, ears, and skeleton, and the acquisition of hair and internal temperature control. (They were, in fact, the first warm-blooded animals.) They split into herbivores and carnivores and expanded their territory into a variety of niches and climates around the world, and by 210 million years ago were giving way to true mammals.

Surviving sauropsids, in contrast, remained in the background until they spawned dinosaurs about 230 million years ago. The oceans were a hospitable home, and sauropsids became quite large sea predators. Eventually, though, some returned to land and became the dominant terrestrial predators. They accounted for most of the large-bodied animals alive during their heyday and spread widely; they were the ancestors of current-day reptiles (alligators, crocodiles, lizards, snakes), as well as birds. Birds are basically flying reptiles with feathers—their hard-shelled eggs tell of their past.

Fast, carnivorous dinosaurs preyed upon mammals, which could not compete in either size or strength. Because dinosaurs were daylight creatures, pressures arose on mammals to become more active at night, and to become smaller and survive on less food. Early mammals were tiny, shrew-like creatures, weighing just an ounce or two, that lived in dense forests, which were hard for dinosaurs to penetrate.

Around 65 million years ago, another mass extinction occurred, in which animals roughly fifty pounds or heavier were wiped out. This spelled the end for most dinosaurs. While it is not certain what led to this mass extinction, a popular theory is that a large meteor produced a cloud of dust upon impact, blocking sunlight and disrupting photosynthesis by plants, and thus altering the food chain. Smaller reptiles, birds, and mammals were impacted less, perhaps because they needed less food to survive.

Christine Janis argues that it was the small size of mammals, resulting from their failure to compete with dinosaurs, that allowed them to survive: "A small animal could be flexible; it could swim, climb, dig, run, or jump as its conditions required. A larger animal had to specialize—and the greater the specialization, the harder it is to alter its body plan to adapt to a changing environment."

The massive continental drift that separated the Earth's landmasses had not yet occurred during the early age of the dinosaurs. This allowed animals native to one area to migrate to others and achieve wide distribution over land. By the time of the extinction of dinosaurs, the continents had begun to spread, and mammals in different parts of the world began to evolve separately and greatly diversify. Further diversification occurred by way of a land bridge that remained between Asia and North America, allowing some further mixing.

Today's mammals include the largest animals on the planet (blue whales can weigh almost two hundred tons and extend for about thirty yards) and also relatively small creatures (the Etruscan shrew is less than two inches long and weighs less than an ounce). Most use legs to locomote on land, but others use fins to swim and still others fly. Some mammals are vegetarians, others carnivores, and still others omnivores. One group, called monotremes, lays eggs (for example, the platypus), but all others give birth to live offspring. There are two such groups. Marsupials (kangaroos) carry their young in a pouch. In placental mammals the fetus remains in the placenta (a modification of the amniotic membranes of reptiles). When the placenta breaks, releasing fluid, it is a sign that birth is near. Most mammals are of the placental variety, including rodents (rats, mice), felines (cats), canines (dogs, wolves), farm animals (horses, cows, goats, sheep, pigs), flying mammals (bats), marine mammals (whales, dolphins, walrus, seals), and primates (monkeys, apes, humans). The various placental mammals fall into four groups, each with a different common placental ancestor, as shown in the mammalian family tree (figure 35.3).

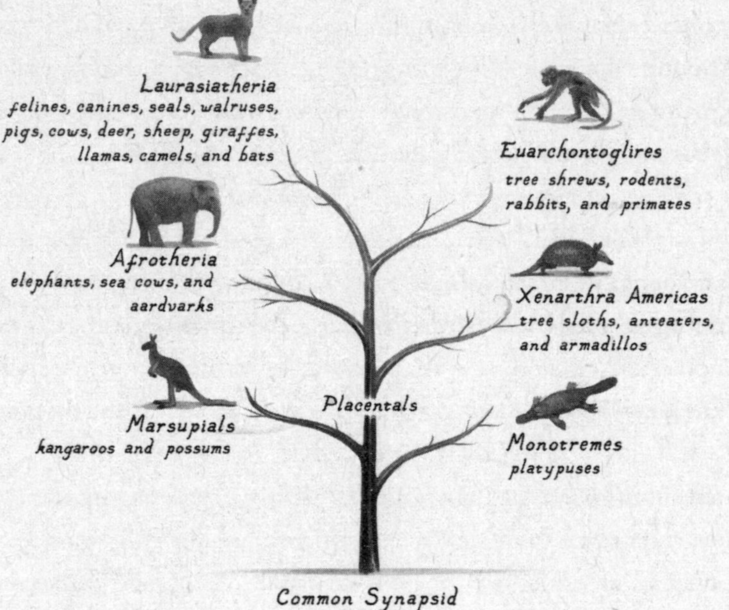


**Figure 35.3:** ***Mammalian Family Tree***

*Chapter 36*

# THE MILK TRAIL

Mammals have a number of features that define their Bauplan and that differentiate them from other vertebrates. They are best known for nursing their young with milk from their mammary glands, a feature that gives mammals their name. Reptiles, including the cynodont ancestors of mammals, continually replace teeth, while mammals are born with no or few teeth, and acquire their first set sometime after birth, and then replace those only once. This situation required a means of nourishing mammalian young, since the babies couldn't chew food. Using food made by the mother (milk), and passed from her to her infants via breastfeeding, was a practical solution, as it also allowed the female to nurture and protect the infants in the home area without having to leave and search for food.

Some of the other signature mammalian traits were already developing in their cynodont, mammal-like, reptilian ancestors. For example, cynodonts had modified the tetrapod skeleton so that their legs would be positioned under their body rather than off to the side. This facilitated breathing while running, which reptiles couldn't do. It also made more oxygen available for metabolism, and internal heating of the body. They were the first warm-blooded animals, or endotherms, able to maintain a core, stable body temperature under variations in temperature of the external environment (up to a point). They probably also had fur, which reduced some of the metabolic burden in the cold. And having converted the reptilian three-chambered heart into one with four, they had a powerful system for delivering oxygen to the tissues that needed to make energy.

Internal temperature control may have also helped cynodonts survive the global cooling that did in cold-blooded dinosaurs, since it allowed more stability in the face of external temperature variations. (Birds independently acquired a warm-blooded body, which may have helped them survive global cooling as well.) Cynodonts also had a double-jointed jaw, out of which unique hearing capacities developed (see below). These features all became part of the mammalian Bauplan.

Being nocturnal, early mammals had less need for color vision. Indeed, they seem to have sacrificed some of the capacity for it that was present in other vertebrates and replaced it with night vision. Biologically, this would have involved swapping color receptors (cones) for light-dark receptors (rods). Of the mammals, only primates have reclaimed complex color vision.

Early mammals also developed expanded capacities for smell, facilitated by their high oxygen intake through the nose, as well as sound-processing enhancements—reptiles and amphibians mainly hear low frequencies; mammals added higher registers. This was made possible, as mentioned above, by the conversion of part of the jaw into a sophisticated ear with an eardrum and middle-ear bones that helped turn sounds into neural messages to send to the brain. This may help explain why some people develop tinnitus (ringing in the ear) from jaw clenching.

Some of these features directly or indirectly required changes in the brain. For example, an HVAC system had to be devised to process external temperature and generate or reduce heat production. Systems for detecting and responding to predators, and some mechanism for mother-infant bonding, was needed. Olfactory and auditory processing systems changed to accommodate the increased importance of these modalities over vision in early mammals, and the visual system had to specialize for night vision.

Our focus here is not, however, on the complete history of mammals per se, but on the history of ourselves. The key modifications we want to examine are thus those that enabled small, nocturnal mammals with poor color vision, leading lives in which they were preyed upon by larger, stronger creatures, to become animals that stood tall on two legs and developed brains that solved problems by thinking and reasoning and that communicated with other brains by speech. This path involved placental mammals that diverged to give rise to small, insect-eating insectivores (tree shrews), and from them to primates (figure 36.1).*

Primates emerged roughly 70 million years ago. Their diet included leaves, flowers, nuts, and fruits, and they had the ability to grasp with their hands and feet, enabling them to leap from branch to branch in trees, and to forage for food high aboveground away from carnivorous mammalian predators. Their eyes moved to the front of their face, making possible

**Figure 36.1: *Primates: From Insectivores to Humans***

* This paragraph is based on Elisabeth Murray, Steven Wise, and Kim Graham, *The Evolution of Memory Systems.*

binocular vision and depth perception, which are important skills for leaping and grasping. They moved mostly by using their legs, rather than on all fours, making the arms available for steering and manipulating items. One hand could be used to stabilize the body, and the other for bringing food to the mouth.

These early primates were called prosimians and included lemurs and tarsiers, among others. Two other groups are anthropoids (monkeys and apes) and hominids (humans). The emergence of anthropoids came with additional changes in their body and lifestyle. They shifted their foraging activities to daytime, which was made possible by the development of a fovea, a region of the retina with a concentration of cone cells that allows high-acuity color vision. Daytime foraging was still riskier because of predation, but with improved vision their ability to detect predators increased. Vision is indeed paramount in anthropoids, as they spend much of their time looking at things. As they diversified, they became bigger, and reverted to movement on all fours, which required more energy. They began to depend on fruits, which provided more energy than leaves, but supply was unpredictable, and there was competition for it with other animals. These pressures led to an increase in brain size, and the addition of new brain areas, enhancing cognitive capacities and making possible survival based on intelligence as much as brawn. Even with these added capacities, though, they were hardly the kings of the planet.

Humans entered the picture some 6 million years ago, competing for existence with other primates and other mammals, which had also undergone considerable diversification. In his bestselling book *Sapiens,* Yuval Noah Harari points out that though early humans stood upright, they were merely another animal alongside the gorillas, monkeys, lions, tigers, birds, fish, and insects. In the food chain, they stood in the middle—preyed upon by larger animals, but predators to smaller ones. There were several different species in the genus *Homo*—our own, *sapiens,* lived among *ergaster, neanderthalensis,* and *erectus.* By ten thousand years ago, other human species had become extinct, and *Homo sapiens,* the last one standing, had begun its dominance over bigger, faster, and more powerful animals.

Harari proposed that a random and perhaps sudden genetic mutation in *sapiens* between seventy thousand and thirty thousand years ago started rewiring the species' brain for a different kind of cognition—one based on abstract thought and language. New evidence has fueled speculation that perhaps both Neanderthals and *sapiens* possessed the ability for symbolic thought and language, and thus the key mutation or mutations may have taken place in the common ancestor of these two species.

Part 8

# Ladders and Trees in the Vertebrate Brain

*Chapter 37*

# NEURO-BAUPLAN VERTEBRATA

Vertebrates are, in evolutionary terms, recent members of the family of life. And while they have defining features that separate them from other metazoans, in order to live day to day they have had to satisfy similar basic survival requirements like all other animals before them—in fact, like all other organisms before them. They have to defend against danger, manage energy supplies and wastes, balance fluids, and reproduce. As we've seen, the way a particular type of animal accomplishes these tasks depends on the kind of Bauplan it possesses.

An animal's Bauplan can be thought of as a composite of the sub-Bauplans that constitute its various tissues, organs, and systems, each of which contributes to the survival potential of the organism. One sub-Bauplan that is especially important to the way its possessor survives is that of its nervous system, which coordinates the body's various other response systems as the organism goes about the business of staying alive and well. As Darwin noted, "Man is constructed on the same general type or model with other mammals. All the bones in his skeleton can be compared with corresponding bones in a monkey, bat, or seal. So it is with his muscles, nerves, blood-vessels and internal viscera. The brain, the most important of all the organs, follows the same law."

Darwin was correct, but didn't take his idea far enough. A common overall Bauplan underlies the nervous system, not just of all mammals, but as we will soon see, also all vertebrates. Vertebrates have certainly added features with the arrival of new species over time, but they have been inserted into the general vertebrate neuro-Bauplan that has been adhered to for hundreds of millions of years, starting with the ancestral vertebrates that diverged from invertebrate chordates.

The canonical vertebrate nervous system consists of two components: the central and peripheral nervous systems. The central nervous system (CNS) is made up of the brain and spinal cord. Neurons in the CNS interact with the outside world by way of the peripheral nervous system, a network of nerves that connects the brain and spinal cord with tissues in the rest of the body. The connections go in both directions, taking information into and out of the CNS: nerves originating in sensory organs (eyes, ears, nose, mouth, skin) transfer information about the external environment to the CNS, while nerves originating in the CNS and terminating in body tissues control responses of the body. One type of response is behavior, which is a product of striated muscles. Physiological responses of internal body systems, like the respiratory, digestive, and circulatory systems, involve smooth and cardiac muscles. Although each of the vertebrate classes and species has sensory and motor specializations tailored to its specific survival needs, these unique features are variations on the common vertebrate sensory-motor Bauplan.

Of particular relevance to our discussion going forward is the Bauplan of the vertebrate brain. This bilateral structure consists of two complementary halves, each preferentially (though not exclusively) responsible for peripheral sensory and motor functions of the opposite side of the body. Within each half brain of all vertebrates are three broad zones: hindbrain, midbrain, and forebrain (figure 37.1). And within each zone are

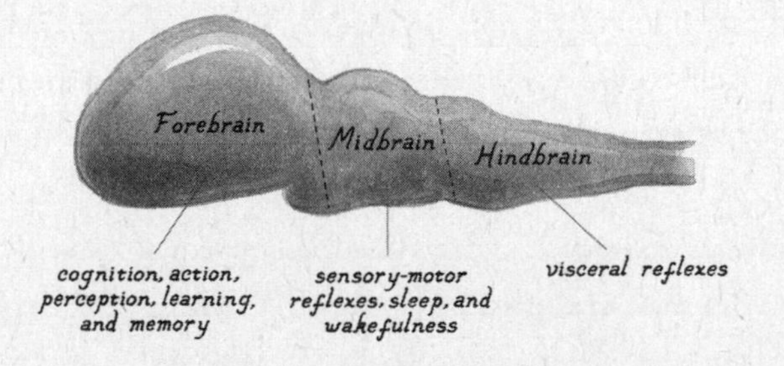

**Figure 37.1:** ***The Three Zones of the Vertebrate Brain and Their Basic Functions***

more specific areas that house the neurons that contribute to specific functions characteristic of the area and zone. At the risk of gross oversimplification, the overall function of the three zones can be caricatured as follows.

Neurons in areas of the hindbrain are responsible for vegetative functions necessary for life, such as those underlying circulatory, digestive, and respiratory activities achieved by smooth and cardiac muscles. Midbrain areas and their neurons are involved in controlling primitive sensory-motor behavioral reflexes expressed through skeletal muscles. Neurons in forebrain areas are responsible for more complex skeletal-motor behaviors, including those based on innate programming and learning, as well as those that depend on cognitive deliberation and decision making, and, in some species, consciousness, including conscious emotions.

Functions are not, strictly speaking, carried out by areas, or even by neurons in areas. They come about by way of circuits that consist of ensembles of neurons in one area that are connected by nerve fibers or axons to ensembles in other areas, forming functional networks. As with other features, the wiring pattern of sensory and motor systems is evolutionarily conserved across the vertebrates.

Any function that involves sensory processing and motor control, which includes all behavioral functions, requires networks that span both the central and peripheral nervous systems. For example, consider the organization of the simple reflex response that causes an animal to orient its head toward a sudden sound. The auditory stimulus travels from the ear to the brain by way of the peripheral auditory nerve bundle. The auditory nerve axons then form synapses on neurons in auditory-processing regions in the hindbrain. These neurons send fibers up to the midbrain, where they connect with auditory processing circuits there. The output nerves of the midbrain circuit then descend to connect with circuits in the upper spinal cord that control neck muscles. The reason that such reflexes are said to be a function of the midbrain is because the midbrain is the highest part of the brain required—in other words, the midbrain is necessary but not sufficient for the response. Similarly, for more complex behaviors—

like those involved in defense, feeding, and reproduction—the forebrain is necessary but not sufficient for the responses, since lower regions in the midbrain, hindbrain, and spinal cord are also involved in connecting the CNS regions with the peripheral nerves that control muscles.

Just as axial patterning along the anterior-posterior axis of the body is controlled by homeobox genes, so, too, is the anterior-posterior patterning of the brain and its circuits. For example, the hox suite of homeobox genes controls the development of the hindbrain and spinal cord, while development of the midbrain and forebrain is controlled by different homeobox gene suites. The central nervous systems of protostomes like fruit flies, bees, and cockroaches also have brains with three major zones that are subject to anterior-posterior patterning by hox genes. This similarity likely reflects a common neural heritage bestowed upon both deuterostomes and protostomes from the PDA and LCBA.

*Chapter 38*

# LUDWIG'S LADDER

Darwinian ideas were the all the rage in the late nineteenth century, prompting biologists to compare the bodies of different animals in an effort to better understand the evolutionary origins of humans. By the early twentieth century, much information was available about the bodies of various vertebrates, but a cohesive picture had not yet emerged, and a synthesis of the findings was sorely needed. The German anatomist Ludwig Edinger did just this for the brain.*

Edinger's efforts yielded an influential observation that shaped thinking about vertebrate brain evolution through much of the twentieth century. He noticed that the hindbrain and midbrain were remarkably similar across different vertebrates, but the forebrain varied in size and complexity over what he saw as a progression from fish through mammals. Edinger proposed that this was due to sequential expansion by layering of new structures in the forebrain as reptiles diverged from amphibians, and again as mammals diverged from reptiles, with further changes continuing in mammals, ultimately resulting in the largest, most complex brain—that of humans.

The human forebrain in Edinger's model, in short, was a mélange of ancestral vertebrate brains—a reptilian, an early mammalian, and a new mammalian minibrain—stacked on top of one another (figure 38.1).

The forebrain consists of two different kinds of brain tissue, often referred to as cortical and subcortical tissues. Key to the Edinger model is

* This discussion of Edinger's theory is based on Georg Striedter's summary in *Principles of Brain Evolution*.

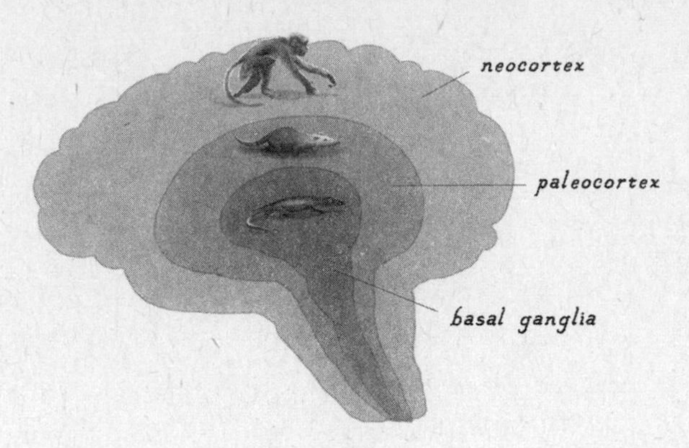


**Figure 38.1:** ***Edinger's Model of Forebrain Evolution***

the sequence in which various specific areas were added to the cortex and subcortex over the course of vertebrate evolution. His assumption was that in fish, amphibians, and reptiles, the forebrain was dominated by the subcortical area known as the basal ganglia, with other subcortical areas and cortex added later in mammals (figure 38.2).

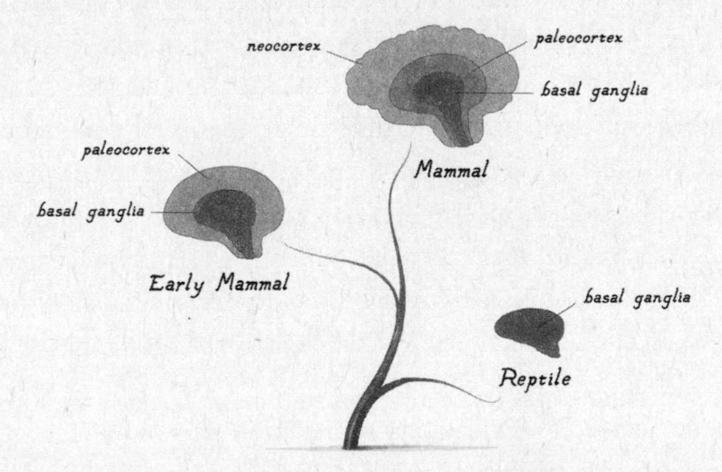


**Figure 38.2:** ***Changes in the Forebrain During Vertebrate Evolution Proposed by Edinger***

Specifically, Edinger proposed that with the evolution of reptiles, the basal ganglia expanded by adding a new subdivision called the striatum. During the evolution of early mammals, still another new region of the striatum was added. Consistent with the proposed evolutionary sequence, the older reptilian part of the basal ganglia was labeled with the prefix *paleo* (e.g., *paleostriatum* means "old striatum"), and the mammalian addition received the *neo* prefix (*neostriatum*). Mammals also added new subcortical areas, the amygdala being a key example. But most significant was the fact that, for the first time, a cerebral cortex appeared.

Collectively, the various additions led to a significant increase in the size of the mammalian forebrain, as compared to that of reptiles. With the continued divergence of mammals over time, even more changes emerged—specifically, an additional, more complex type of cortex called the neocortex. It was expansive, and increased the mammalian forebrain even further, and resulted in the older cortex coming to be known as the paleocortex. The neocortex is the wrinkled mass often illustrated in pictures of the brain. The paleocortex, by contrast, is located medially, and not visible unless the two hemispheres of the brain are pulled apart (figure 38.3).

Starting in the 1970s, new research methods for studying the brain were developed, spawning a fresh wave of neuroevolutionary findings.

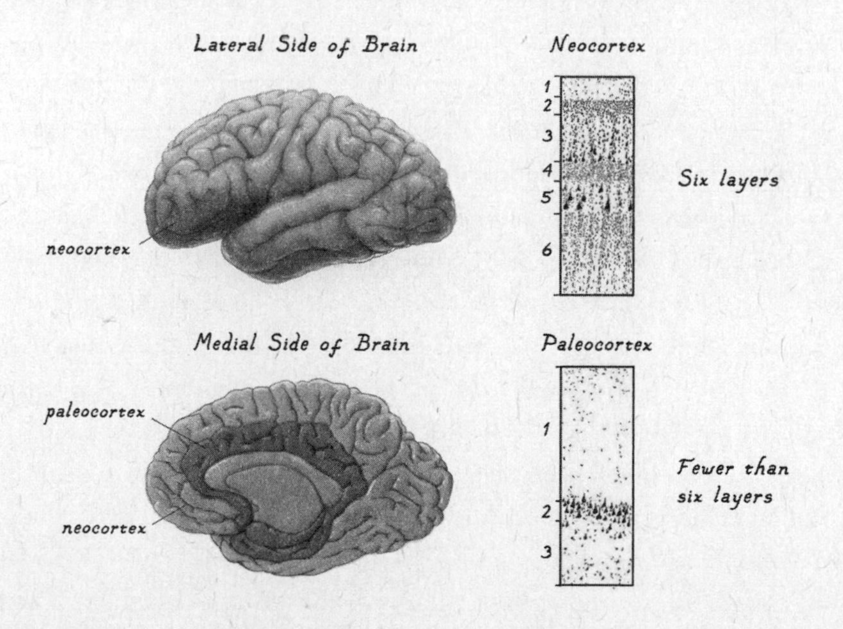


**Figure 38.3:** ***Neocortex vs. Paleocortex***

Evidence obtained by Harvey Karten, Glenn Northcutt, and others strongly challenged the basic premise of Edinger's sequential building-block theory. Most striking was the discovery that homologues of the neocortex exist not only in present-day representatives of early mammals (which should have only had a paleocortex), but also in reptiles and birds (which, being nonmammals, should have neither paleocortex nor neocortex).

Other key parts of the sequential theory also proved wrong. Edinger argued that amphibians and fish have basal ganglia but not the striatal part, which, he said, only appeared with reptiles. But more recent findings show that fish and amphibians have a striatum as well. Moreover, the basal ganglia does not account for as much of the anterior forebrain in reptiles as Edinger believed, since the reptile forebrain actually has a paleocortex, amygdala, and septum, all of which were supposed to only have arrived with the mammals. Today, as we learned in the previous chapter, all vertebrate brains are thought to follow a common general Bauplan, with differences being mainly in degree, rather than kind.

Edinger strove to achieve a Darwinian account of brain evolution. While his model was true to the spirit, it was not true to the letter of Darwinian theory. As noted in chapter 2, Darwin argued that the history of life is more accurately portrayed in terms of a branching tree metaphor than by the ladderlike, linearly progressing *scala naturae* concept originated by Aristotle and expanded upon in Christian theology. The idea that completely new Bauplans (bodies) and sub-Bauplans (nervous system parts and functions) arrived sequentially across vertebrates, eventually culminating in the human brain, is in keeping with the ladder metaphor.

The designations *neo* and *paleo** in the ladder model came to imply not only age but also quality. Newer areas came to be thought of as superior to older ones, which were considered less evolved. But as Ann Butler and William Hodos note, evolution does not create superior or inferior organ-

---

* Some contemporary neuroscientists prefer to avoid labels such as *neo* and *paleo* to circumvent neuroevolutionary implications. For example, they refer to layer 6 cortex as isocortex and paleocortex as allocortex. While I support this, I reluctantly adhered to the more conventional terms here, saving my semantic battles for the language of emotion.

isms or tissues; it creates diversity through divergence (not by accumulation of features). What works in a given environmental situation is determined by natural selection, but as the environment changes, or the group moves to a new niche, new traits become important and previously useful traits can become detriments.

In trying to understand how changes in brains over the course of vertebrate evolution gave rise to our own behavioral and mental capacities, we thus have to take care to climb the vertebrate tree of life without using Edinger's ladder, and also to be cautious not to overinterpret the differences we encounter. We are newer and different, but not better. As I have pointed out, we have to guard not only against anthropocentric tendencies, but also against anthropomorphic ones. In other words, we sometimes attribute too much to other animals, and sometimes to ourselves. We have to find the right balance, which I strive to do in the coming chapters.

*Chapter 39*

# THE TRIUNE TEMPTRESS

One of the most influential developments in early brain research was Paul MacLean's synthesis of findings in the 1950s regarding brain evolution in relation to emotions. The backlash against Edinger was still two decades away, and MacLean, like most scientists at the time, accepted the validity of Edinger's tripartite scheme of hierarchically arranged layers.

Fascinated by psychosomatic diseases, which at the time included conditions like high blood pressure and stomach ulcers, MacLean sought to discover how troubling thoughts and ideas—basically, stress and anxiety—could cause body dysfunction. In the 1920s, the physiologist Walter Cannon had shown the importance of the hypothalamus in regulating body functions associated with hunger, thirst, and rage. MacLean speculated that mental states occurring in the cortex were somehow disrupting this hypothalamic regulation of body physiology. The problem was that mental states were believed to be products of the neocortex, which had no connections to the hypothalamus. However, the paleocortex, the functions of which were then not well understood, did have connections to the hypothalamus, both directly and by way of the subcortical forebrain areas like the amygdala and septum. MacLean put his eggs in the paleocortical basket.

MacLean took Edinger to heart and described the mammalian forebrain as consisting of three conserved evolutionary partitions—the reptilian, paleomammalian, and neomammalian complexes. He introduced the more poetic term, *triune brain,* and proposed that each of the three parts has an evolutionarily based specialized function. He thereby attempted to do for the evolution of brain function (behavior) what Edinger had attempted for the evolution of brain structure.

In MacLean's triune brain theory the reptilian complex, essentially the basal ganglia (with some midbrain and hindbrain extensions), was responsible for primitive, species-specific behaviors—the raw, instinctive animal reactions that are automatically triggered, such as aggression, dominance, and territoriality.

The paleomammalian complex, in contrast, consisted of the older cortex (paleocortex) and interconnected subcortical forebrain areas (amygdala, septum). The key paleocortical areas included the hippocampus and cingulate cortex. Centuries earlier, this cortex was noted to form a rim (limbus) in the medial wall of the hemisphere. This is why MacLean famously designated the paleocortex as limbic cortex, and dubbed the entire paleomammalian complex as the limbic system (figure 39.1).

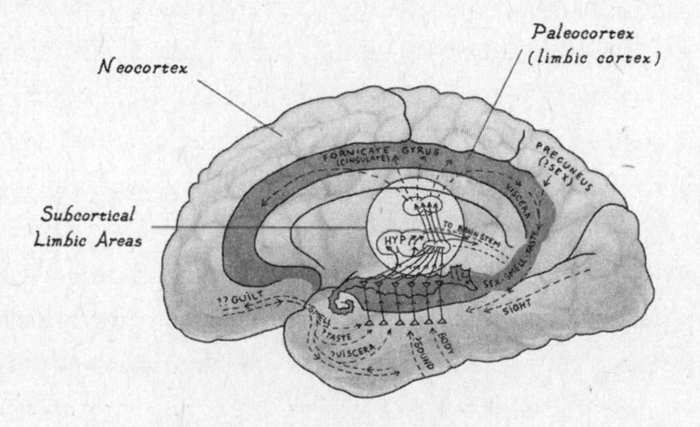


**Figure 39.1: *The Limbic System***

MacLean proposed that the paleomammalian complex made possible, for the first time, emotions. These inner conscious states were associated with so-called emotional behaviors related to feeding, defense, and reproduction. Because emotions are imbued with positive or negative affect, their occurrence in daily life reinforces the learning of new responses that occur in the service of survival needs. Through such reinforcement learning, behavior became more flexible in mammals relative to the rigid, precedent-driven, instinct-bound reactions of reptiles.

Underlying these novel functions were connections of the paleocortex to the hypothalamus. By way of these connections, and indirect connections via the amygdala and septum, emotions formed by cortical and subcortical centers of the limbic system controlled behavioral and visceral functions of the body.

At the time, Freud's theory of unconscious emotion was very popular. Freud had actually initially hoped to relate the mind to the brain but concluded that the brain was not sufficiently understood, and instead he developed a psychological model. MacLean tried to unite Freudian psychology with the brain, proposing that the limbic system was the home of unconscious emotion, essentially the Freudian id, and the source of pathology in psychosomatic disease and other psychiatric problems.

While the paleomammalian limbic brain, according to MacLean, could make associations between stimuli and responses via reinforcement, it had limited ability to analyze and reflect on them. This ability was proposed to have arrived in later mammals with the evolution of the neomammalian complex. This complex is one and the same as the neocortex, which, according to MacLean's theory, made possible complex cognitive capacities such as those underlying the ability of late mammals to think, remember, plan, and decide, and, in humans, to talk. Just as the Freudian id was housed in the limbic system, the Freudian ego was part of the thinking, speaking neocortex.

The limbic system/triune brain theory is still immensely influential in scientific and lay circles. But it remains problematic. First, it inherited all the anatomical problems of the Edinger model: the "neocortex" did not first appear in late mammals; the "paleocortex" and subcortical limbic areas did not first appear in early mammals; and the basal ganglia did not appear in reptiles. Second, the sequence of arrival of behavioral capacities did not occur in the way proposed. For example, MacLean's distinction between instinctual and emotional behaviors was in part based on Edinger's assumptions about the age of the basal ganglia and limbic system, and by his psychological theory about how these anatomical changes enabled mammals to have emotions. But what he called instincts (behaviors related to aggression, dominance, and territoriality) and emotional behaviors (those related to feeding, defensive, and reproductive activities) do occur in reptiles and birds, not just in mammals. Such species-typical survival behaviors exist to keep the organism alive. The fact that, in some species, notably humans, these kinds of behaviors are associated with certain emotional feelings does not mean that feelings arose to control the innate behaviors.

Finally, the limbic-system theory was based on two interrelated and flawed assumptions about human brain function: that areas of the limbic system are involved in emotion but not cognition, and that areas of the neocortex underlie cognition but not emotion. Both are wrong. Areas of the paleocortical limbic system, like the hippocampus and cingulate cortex, contribute in significant ways to cognitive functions like memory and attention, and neocortical areas contribute extensively to felt emotional experiences.

Edinger and MacLean were both pioneering thinkers and researchers, and did a remarkable job formulating ideas on the basis of what was known in their day. They stimulated a tremendous amount of research. While evolution does not progress toward perfection, science tries to get closer and closer to perfect answers to its questions, and sometimes ideas that pushed the field forward in the past, no matter how intuitively compelling, have to be pushed aside as the knowledge unfolds.

*Chapter 40*

# DARWIN'S MUDDLED EMOTIONAL PSYCHOLOGY

Our emotions are central to our mental life. Figuring out how they work in the brain is thus of great interest, and an important goal because of their role in mental suffering. But our ability to understand how any psychological process works in the brain is only as good as our understanding of the process itself. In other words, we have to know what we are looking for in order to find it. I believe that brain researchers have, for the most part, misunderstood what emotions are, and have searched for them in the brain in the wrong way. Edinger's theory of brain evolution, as we've seen, suffered from ignoring Darwin's biological insight that the history of life is more like a branching tree than a ladder. Theories of emotion, and of the emotional brain, in contrast, have suffered because they adopted Darwin's psychological ideas.

Darwin, like most scholars at the time, assumed that words like *mind* and *mental* referred to consciousness, a view strongly influenced by the seventeenth-century philosopher René Descartes. Descartes considered animals to be thoughtless brutes, so-called beast machines that simply reacted to their environment, being pulled and pushed by stimuli in daily life. Humans, however, had conscious minds that made possible the experience of inner states and the use of these experience states in the control of behavior.

With Darwin, humans came to be seen, for the first time, as part of a continuum with other animals. To use Desmond Morris's famous phrase, we are "naked apes." But Darwin's concept of psychology, rather than explaining human qualities on the basis of animal traits, called upon human psychological features, especially mental states, such as emotions, to explain the behavior of other animals. Elizabeth Knoll argues that he chose to take this approach in part because he was trying to get his ideas about continuity to be accepted in Victorian England, a time in which anthropomorphic sentiments were popular among the upper middle class. Darwin himself noted that arguing for humanlike traits in animals, rather than animal-like traits in humans, allowed him to make his point about animal-human continuity in a more "cheerful" way.*

Of particular interest are the views in Darwin's 1872 book, *The Expression of the Emotions in Man and Animals.* In it he argued that emotions are states of mind that humans have inherited from our mammalian ancestors because these mental states helped our ancestors adapt, survive, and reproduce. Darwin realized that we don't have direct information about an animal's mind, but he assumed that behavioral responses are a direct reflection of what an animal, including a human, feels. If a monkey or dog responds the way a human does when threatened, the animal must, like a human, feel fear. In fact, fear must be the cause of the behavior. The book was filled with illustrations showing similarities in facial expression and bodily postures between humans and other mammals. And he wrote freely about the mental underpinnings of the responses. His disciple, George Romanes, continued in Darwin's footsteps when he referred to behavior as the "ambassador of the mind," and meant this to apply to animal and human minds alike. Many followed their lead, freely interpreting animal behavior in terms of conscious states comparable to those experienced by humans. These excesses led to a backlash in the early twentieth century—the behaviorist revolt in psychology.

In Darwin's defense, the idea that there is more to the mind than meets the conscious mind's eye was not recognized at the time. Nonconscious mental processes did not enter into the scientific discourse for some decades, initially through the writings of Freud, and later via the arrival of cognitive science (which we'll explore in the next section). Had Darwin

---

* The literary genre of animal autobiography, which features animals as narrators of their own stories, including experiences with human cruelty, was thriving—*Black Beauty,* the mistreated talking horse, was a blockbuster.

been privy to such information, he may have taken a different approach to the nature of psychological continuity between humans and other animals. This would have changed the direction that emotion research has taken since. But he took the course he did, and we are living with the consequences in the study of emotion.

*Chapter 41*

# HOW BASIC ARE BASIC EMOTIONS?*

William James, the father of American psychology, was generally a fan of Darwin's ideas, but not of the commonsense approach to emotions so important to Darwin. James famously noted, "Our natural way of thinking about . . . emotions is that the mental perception of some fact excites the mental affection called emotion, and that this latter state of mind gives rise to the bodily expression." He then went on to reject the idea that emotions, or feelings, are mental states that cause us to behave in certain ways, stating that we do not run from a bear because we feel afraid. Instead, he proposed, we are afraid because we run—that the act of responding to danger generates physiological signals that are interpreted as fear. James's rejection of the idea that fear is why we run from the bear is often overshadowed by his explanation of where fear comes from. In other words, he was less interested in why we run than how we feel, and tended to emphasize the role of the body's responses in causing feelings.

The new field of psychology, which James helped to establish, had emerged in an effort to use experimental methods to understand Descartes's problem of consciousness. But psychology's focus on consciousness was short-lived, as the behaviorists soon initiated their hostile takeover and eliminated all things mental from the field.

Meanwhile, research on the emotions was thriving in departments of physiology, which were often based in medical schools, where the behaviorists' ideas had little impact. Early pioneers in this effort seamlessly con-

* This title is a nod to Andrew Ortony and Terence J. Turner's "What's Basic About Basic Emotions?"

tinued Darwin's anthropomorphic tradition, freely calling upon inner experiences of fear, rage, hunger, or pleasure to explain the role of brain areas as the cause of behavioral responses in animal subjects. Their findings and conclusions provided the intellectual tradition out of which MacLean's limbic system theory emerged after World War II, which is why it was so natural for him to assume that an animal's behavioral responses revealed emotional states encoded in the limbic system.

By the 1960s, the behaviorists' influence in psychology had begun to wane, and a new wave of interest in emotions as mental states arose in parallel with the cognitive revolution. Some, riding the tide of that revolution, emphasized the role of cognition in the creation of feelings—namely, that an emotion like fear is the result of the interpretation that one is in a situation of danger (this will figure prominently in later discussions). Others, however, revived Darwinian ideas about the innateness of emotions.

Silvan Tomkins, for example, proposed that the human brain has several biologically based basic emotions, which have been inherited from animal ancestors, and which are expressed in our facial expressions and other bodily responses. These included fear, anguish, shame, disgust, rage, joy, interest, and surprise. Psychologists like Robert Plutchik, Carroll Izard, and Paul Ekman took up the basic-emotions cause and used the seeming universality of facial expressions of emotions around the world as evidence. Ekman's ideas have been particularly influential in psychology and neuroscience, but also reached far beyond scientific circles: his theory was the basis for the popular film *Inside Out*.

Basic-emotions adherents propose that each emotion is bundled within an affect program, a hypothetical neural module that, when activated by an appropriate stimulus, unleashes behavioral and physiological responses and corresponding feelings (figure 41.1).* These theorists didn't have much to say about how the various discrete affect programs might be differentially represented in the brain. If they mentioned the brain at all, it was typically

* These researchers were for the most part more interested in the innate body responses than subjective experiences. Nevertheless, they were postulating theories of emotions, rather than theories of emotional responses, and it was often implied that the subjective experience of emotion was also a product of the affect program.

with reference to the limbic system, as this was the dominant view of the emotional brain at the time.

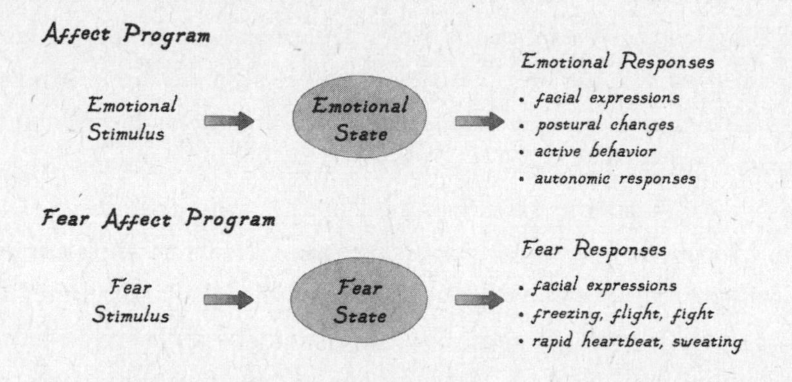


**Figure 41.1:** ***The Affect Programs of Basic Emotions Theory***

MacLean himself did not say much about how specific emotions arise, as he mainly focused on the limbic system as a general-purpose circuit of emotions. But the neuroscientist Jaak Panksepp later picked up where MacLean left off, developing a basic emotions theory in the brain.

Panksepp identified distinct emotion circuits—what he called emotion operating systems (another name for affect programs)—by stimulating limbic areas in rats and eliciting what he termed emotional behaviors. Each emotion operating system, Panksepp argued, generates both the characteristic behavior and the core feeling of its evolutionarily designated emotion. In real life, as opposed to a brain stimulation experiment in the laboratory, when danger is encountered, "fear," he said, is aroused in the circuit that also controls so-called fear responses, such as freezing or flight. Because these operating system circuits are responsible for both feelings and behaviors, and are highly conserved across mammals, Panksepp assumed that the circuits responsible for human emotional feelings could be revealed by identifying the circuits that control emotional behaviors in other mammals.

But I believe that this mischaracterizes what his so-called subcortical operating systems actually do. For example, Panksepp's fear operating system involves the amygdala and other subcortical areas connected with it.

Research that I and others have done has indeed implicated the amygdala in the detection of and response to danger. For example, rats, mice, cats, monkeys, and humans that have sustained damage to the amygdala fail to exhibit typical behavioral and physiological responses when they encounter threats to their well-being. And amygdala neurons are activated when animals or humans are exposed to threatening stimuli. But does this mean that the amygdala is responsible for the *feeling* of fear? Panksepp argued this is the case, but the evidence he used is purely behavioral. He followed the Darwinian logic—since humans experience emotions when they are behaving in these ways, when an animal behaves like this it must be feeling what we feel. Studies measuring behavior in animals can thus tell us where feelings live in the brain of animals and humans. But I say, not so fast.

When threats are presented subliminally to human participants, the subject's heart beats faster, his palms sweat, and his muscles tense, but he's not aware of the stimulus and doesn't report any feeling of fear. Conversely, a person with amygdala damage can actually report feeling fear, despite being deficient in generating the bodily responses to threats. These findings are more consistent with a model in which the amygdala is responsible for the detection of and the initiation of response to danger nonconsciously, but is not itself directly responsible for the conscious feeling of fear that occurs at the same time. The nature of this distinction will be discussed in detail when we turn to contemporary research on the nature of emotion in the last part of the book.

The above discussion should not be taken to mean that the amygdala has no role in fear. The behavioral and physiological responses triggered in the body by amygdala activation feed back to the brain and do, in fact, affect how we experience the dangerous situation. For example, hormones released in the body travel in the bloodstream and enter the brain, helping focus attention and amplify our experiences. But we can also cognitively notice when we are freezing or fleeing, and when our heart is pounding, and those self-observations also affect what one experiences in such situations. Moreover, the amygdala has connections with numerous other brain regions and has significant effects on information processing in the brain, including on the processing that I hypothesize is crucially involved in the assembly of the actual conscious experience of fear. And a similar scenario applies to circuits that control behaviors related to feeding, fluid balance, thermoregulation, reproduction, and other life-sustaining activities; they are not the sources of the actual feelings of hunger, thirst, and pleasure, but typically do make important contributions to such experiences.

The question of whether animals have emotional experiences has thus become conflated with the question of whether innate behaviors are the way to measure emotions, and to find emotions in the brain. How emotional experiences come about, and the role of life-sustaining circuits in such experiences, will be discussed once we have explored the nature of consciousness in the coming parts of the book.

The basic-emotions approach helped focus attention on the contribution of subcortical circuits to the control of innate responses that are shared across species. And it showed how responses like facial expressions are fairly universal in humans. But I believe it is limited as an account of human feelings. It is interesting that Paul Ekman, a leader of human basic-emotions research, has been much more interested in facial expressions and has been rather noncommittal about the role of affect programs in feelings. Many others, though, write and talk as if they believe that feelings are producers of the innate circuits.

Andrea Scarantino, a basic-emotions theorist, has recently made an effort to reconcile basic-emotions theory with my views. But his approach leaves intact the idea that fear has conscious and nonconscious components, which, in my opinion, will always be problematic.

That the way we use words matters is made acute by the growing number of scientists who wholeheartedly embrace anthropomorphism, claiming that animals have emotions and other states of consciousness comparable to those that humans experience. Prominent examples, other than Panksepp, include Frans de Waal, Gordon Burkhardt, and Marc Bekoff. Each has argued for a "scientific anthropomorphism," using terms like *critical anthropomorphism*, *biocentral anthropomorphism*, *animal-centered anthropomorphism*, or *zoomorphism*.

Many are challenging these ideas, arguing that versions of anthropomorphism with scientific-sounding names do not qualify as rigorous scientific approaches to behavior. Cecilia Heyes, for example, has noted that too often claims are made on the basis of anecdotes and simple analogy with human behavior—that if animals act in ways similar to the ways humans do in similar situations, they must have the same feelings. Unless one can rule out alternative nonconscious interpretations in animals, claims of consciousness should be withheld.

Interestingly, Frans de Waal argues that "gratuitous anthropomorphism is distinctly unhelpful." That's an important qualification, though he also notes that "experienced observers" should be trusted to make the call and be allowed to speculate about the emotions of animals. In my opinion, there is no way, even for experienced observers, to rule out nonconscious explanation by simply watching behavior. Speculation is fine, but when emotions like empathy, joy, and love are spoken about as facts, gratuitous anthropomorphism inevitably results. I'll also have more to say about anthropomorphism later in the book.

Recent trends in basic emotions include approaches in which affect programs continue to contribute to emotion but in a less restrictive way. For example, James Coan treats emotions not as subjective experiences, but as emergent states that include amygdala activity, feedback from behavioral and physiological responses, and subjective experience. In other words, the experience is not the emotion but instead a factor that contributes to the emotion. This is an interesting approach, but I disagree with Coan's treatment of subjective experience as just another factor in a mix of others that make emotions.

For me, the subjective experience—the feeling—*is* the emotion. These are not hardwired states programmed into subcortical circuits by natural selection, but rather cognitive evaluations of situations that affect personal wellbeing. They thus require complex cognitive processes and self-awareness. Much of the rest of this book is concerned with the roots and origins of human cognition and consciousness. This will set the stage for me, at the very end, to present my cognitive view of emotions as conscious experiences.

Part 9

# The Beginning of Cognition

*Chapter 42*

# COGITATION

Our species has survived and thrived not by being bigger, faster, or stronger, but by being clever. We haven't, like most organisms, simply evolved by adapting our Bauplan to the world as it changes; we have used our cognitive abilities to change the world. We do so because we think it might be advantageous to our bodies and way of life, or that it might simply be interesting to tinker with nature. No other animal, not even our closest primate relatives, can have an idea like building a skyscraper, finding a cure for a disease, composing an opera, or writing a novel, then describe it to a colleague, plan how to execute it, and carry it out. That human cognition is unique in no way means we are better or more entitled than our ancestors or animals with which we currently share the planet. It just means we are different.

Unique though it may be, human cognition emerged by building on cognitive capacities possessed by our mammalian ancestors. To understand the origin of our cognitive capacities, we must first specify what they are.

Most commonly, cognition is used in relation to thinking, reasoning, planning, deciding, and the like. Cognition has been a key part of philosophical understanding of human nature since the ancient Greeks. But it was Descartes's famous dictum, *Cogito, ergo sum,* "I think, therefore I am," that equated cognition with a self-reflective consciousness, or an inner awareness of one's self as an integral part of the thinking experience. In Descartes's view, consciousness is a defining feature of what it is to be human, and, as we've seen, for him animals were mindless reflex machines.

A couple of centuries later, Darwin endowed beast machines with humanlike thoughts and emotions, and his anthropomorphism, as noted, ultimately led to the behaviorist revolution in psychology. Like Darwin, behaviorists wanted to narrow the psychological gap between humans and animals, but they took a completely different approach, eliminating consciousness as a factor in both human and animal behavior. Typifying this sentiment, the behaviorist philosopher Gilbert Ryle dismissively referred to consciousness as "the ghost in the machine."

The behaviorists took behavioral continuity to the extreme, turning it into equivalence. They argued that the principles of behavior are universal—all of what one needs to know scientifically about human behavior, including language and thought, could be discovered from laboratory studies of animals. Nothing inside the head, the black box, was relevant, including the brain (figure 42.1).

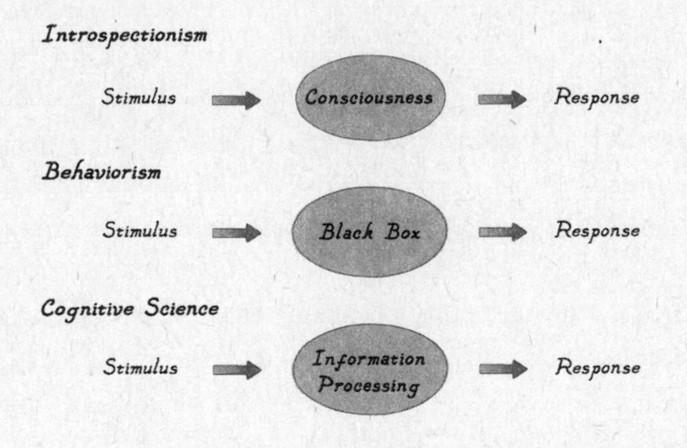


**Figure 42.1:** ***Introspection, Behaviorism, and Cognitive Science***

Based on the apparent similarities between the way humans think and computers process information, a new approach to psychology began to emerge around the middle of the twentieth century. "Cognitive science," like behaviorism, took a rigorous approach focused on behavioral responses. While it used behavior to measure inner states, it eased into these by initially sidestepping consciousness. The cognitive mind was viewed as an information-processing system. And cognitive processing could sometimes result in conscious experiences, but the focus was on the processing, which, in effect, was treated as nonconscious (that is, unconscious) activity. Cognitive science thus brought the mind back to psychology, but not quite the conscious mind that behaviorists had eliminated.

Sigmund Freud had popularized the notion of the unconscious long before cognitive science emerged. His unconscious was a place where troubling conscious thoughts and memories were shipped and isolated to prevent them from provoking feelings of anxiety. But cognitive science offered a different view. The so-called cognitive unconscious, as nonconscious cognitive processes came to be known, is not unconscious because of information repression, but instead because it is organized in such a way that much information processing simply takes place outside of the province of consciousness.

For example, when walking toward a destination (say going to the deli across the street from your office to get a coffee), once you decide to do so, you don't have to think about how to get there—you just go. Similarly, when you speak you usually do a decent job of generating grammatical sentences without having to consciously plan the placement of the parts of speech. This allows you to consciously think about other things while the routine work is being carried out in the background. But if something goes wrong while on automatic pilot (there is an obstruction along the path to the deli, or a sentence doesn't come out right), you take notice. That is, unexpected or undesirable events grab our attention, making their presence known as conscious content, moving out whatever else we were thinking about at the time. Control processes of the cognitive unconscious thus not only underlie the information content that we consciously experience but also direct behavioral interactions with the environment. To distinguish the cognitive from the Freudian unconscious, I prefer using the term *nonconscious.*

If we are going to explore cognition from an evolutionary point of view, we need a precise definition of what it is. As used here, cognition will refer to processes that underlie the acquisition of knowledge by creating internal representations of external events and storing them as memories that can later be used in thinking, reminiscing, and musing, and when behaving. Its dependence on internal representations of things or events, in the absence of the external referent of the representation, is what makes cognition different from noncognitive forms of information processing. Given this definition, processes that allow behavioral responses to an immediately present stimulus are not, strictly speaking, under cognitive control. Only responses that depend on internal representations are. As we'll see, because cognitive science figured out how to make such distinctions, it thrived.

One of the offshoots of cognitive science, and its early connection to computer science, was the field of artificial intelligence (AI). Some argue for what is called strong AI—the idea that cognition, and even consciousness, can result from information representation in artificial systems. I tend to go with a weaker version—that similarities between human cognition and information processing in artificial systems can be valuably exploited for research purposes to help understand human cognition. In other words, the flow of electrons in electronic devices can shed light on cognition but is not sufficient to create it.

My view is thus that cognition is a product of biological evolution, and as such, requires biological information processing. Still, not all instances of biological information processing count. Every cell, for example, engages in biological information processing at every moment of its life. Some scientists restrict cognition to biological information processing underlying behavior. According to this position, behavioral activities of plants, fungi, and even of unicellular microbes reflect rudimentary cognitive capacities. In his book *The First Minds*, Arthur Reber claims that because bacteria exhibit phototaxic responses, they have cognitive minds. In my opinion, the equation of cognition with the ability to generate a response to environmental stimulation stretches the term so far as to make it meaningless.

The view that I will be pursuing is that cognition is a product of biological processes made possible by a nervous system (figure 42.2). This means that cognition is a feature that only evolved in animals, and only in some of them. Given how I have defined cognition, the job of determining which animals are cognitive creatures, and which are not, is thus one of determining which animals have nervous systems that can form, store, and use internal representations.

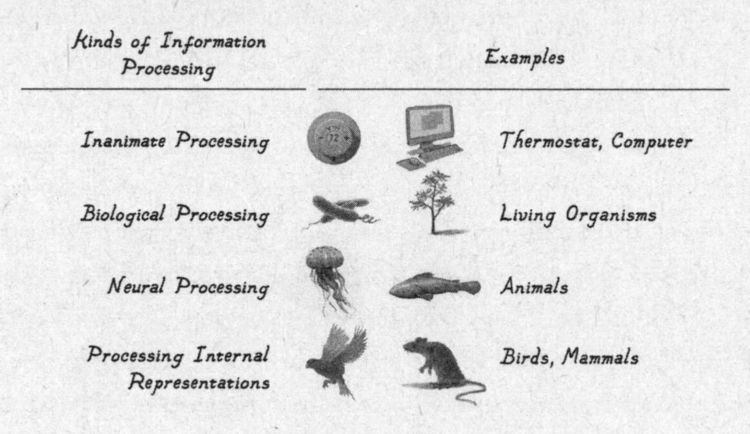


**Figure 42.2:** ***Varieties of Information Processing***

*Chapter 43*

# FINDING COGNITION IN THE BEHAVIORIST BAILIWICK

The philosopher Immanuel Kant argued that one thing science would never be able to do is measure the mind. But hundreds of years later, measuring the mind is precisely what cognitive science figured out how to do—not only in humans, but also in animals.

To measure cognition you have to measure behavior, and, as we've seen, you have to distinguish cognitive from noncognitive behavioral control. Cognitive scientists did this by extending the rigorous approach established by the behaviorists.

The workhorses of behavioral measurements in animals are, of course, Pavlovian and instrumental conditioning, which we explored in part 2. While John Watson founded behaviorism on the Pavlovian procedure, his heir, B. F. Skinner, believed that Pavlovian conditioning was too limited to account for complex human behavior, and turned to Edward Thorndike's instrumental conditioning approach.

To explain conditioned responses, behaviorists examined the subject's history with stimuli and responses. Pavlovian conditioning was accounted for in terms of associations formed by the contiguous occurrence of two or more stimuli, and instrumental conditioning by the contiguous occurrence of responses and reinforcing stimulus outcomes they produced. These contiguous relations came to be called mere associations,* to make it clear that the term *association* did not refer to any internal mental process. Pavlovian mere associations are depicted in table 43.1.

---

* This expression reflects the influence on the behaviorists of the seventeenth-century philosopher David Hume, who had written about the role of "mere contiguity" in learning.

TABLE 43.1: Pavlovian Mere Association (CS and US occur contiguously)

| PAVLOVIAN TRAINING | ABBREVIATIONS |
|---|---|
| CS (light) + US (food) | **CS**—conditioned stimulus |
| PAVLOVIAN TEST | **US**—unconditioned stimulus |
| CS (light) → CR (approach food) | **CR**—conditioned response |

Pavlovian conditioning, it seems, is a widespread capacity among animals. It is present in all classes of vertebrates, as well as in many bilateral invertebrates, including deuterostomes (chordates and echinoderms) and protostomes (insects, mollusks, arachnids), and even in radial cnidarians (jellyfish). Some evidence also exists that it is a capability of plants, protozoa, and even bacteria; unlike animals, these organisms do not have neurons and nervous systems, and thus cannot use synaptic plasticity between neurons to form associations. Instead they use molecular changes within their cell or cells to store the association.

In the early 1960s, findings began to appear that could not be explainable in terms of mere association. The first came from an odd corner of the field. During the height of the Cold War, when nuclear confrontations were a looming threat, the psychologist John Garcia began studying the effects of radiation on behavior. Noticing that rats avoided drinking from water bottles in a chamber in which they were "radiated," he designed an experiment to figure out why. Garcia capitalized on the fact that rats prefer to consume sweetened over plain tap water. But when Garcia treated rats with radiation after consuming sweetened water, they began to avoid it. He hypothesized that the rats avoided the sweet taste because they had associated it with nausea sensations caused by the radiation. He then supported this idea by doing the study with a nausea-producing chemical rather than radiation.

The "Garcia effect" became a new paradigm, called conditioned taste aversion, for studying learning, and it transformed the field. Since hours occurred between the conditioned stimulus (taste) and the arrival of the chemical that produced the unconditioned stimulus (nausea sensations), contiguity could not be involved. What was learned, instead, was an association between the nausea sensations and a memory representation of the earlier taste; the memory of this association allowed the animals to avoid harm.

Memory was one of those inner processes banned in behaviorists' explanations—they preferred to talk about learned behaviors rather than remembered internal states. Garcia's experiment was so revolutionary because it showed that under some conditions Pavlovian conditioning could depend on cognition—on internal representations (table 43.2).

TABLE 43.2: Pavlovian Cognition
(aka Garcia effect, conditioned taste aversion)*

**1. PAVLOVIAN TRAINING**

CS (taste) + delayed US (nausea)

**2. PAVLOVIAN TEST**

CS (taste) → CR (avoid CS)

ABBREVIATIONS

**CS**—conditioned stimulus

**US**—unconditioned stimulus

**CR**—conditioned response

*CS and US are separated in time and the association depends on memory of CS.

Decades earlier, Edward Tolman had proposed a purposive behaviorism, which held that inner organismic factors called intervening variables mediated between stimuli and responses when animals and humans behave in the world (figure 43.1). These variables were viewed as psychological factors, but not necessarily conscious states, which allowed organisms to have inner purpose, while, at the same time, avoiding Ryle's "ghosts in the machine." Particularly radical was Tolman's notion of cognitive maps, psychological models of the world that organisms use to guide behavior. But Tolman's views remained a side story in psychology for decades until

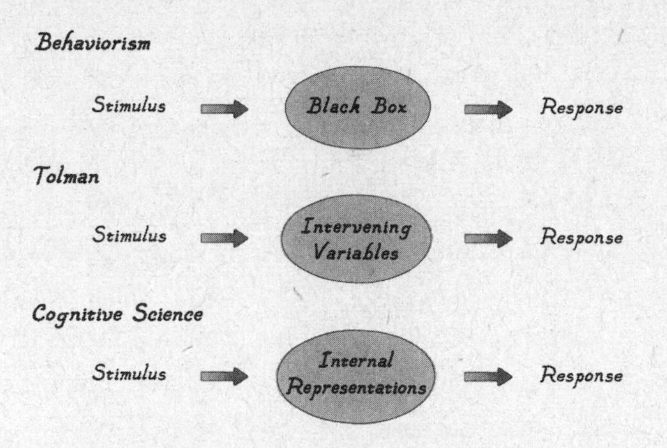


**Figure 43.1:** ***Tolman's Intervening Variables Approach to Behaviorism Anticipated Cognitive Science***

he came to be recognized as an important forebearer of the cognitive revolution.

One key factor in the revival of cognitive maps was John O'Keefe's discovery in the 1970s that cells in the hippocampus (a key part of the brain's memory system) created representations (cognitive maps) of the spatial environment on the basis of the relation between various fixed cues that served as orientation anchors. O'Keefe called these place cells and proposed that they are recruited to form a Tolman-like cognitive map of space as the animal moves around in the world, navigating in search of food, drink, and mates, and in avoiding danger. O'Keefe shared the Nobel Prize in 2016 for this pioneering research.

A key question for us is, Which animals have the cognitive capacities necessary to use internal representations in Pavlovian conditioning? Evidence from unicellular organisms and radial animals is more consistent with the mere association account. However, studies of protostome invertebrates and a variety of vertebrates suggest they can form Pavlovian associations by both mere contiguity and by using simple internal representations (figure 43.2).

Is this capacity shared by vertebrates and protostomes because both

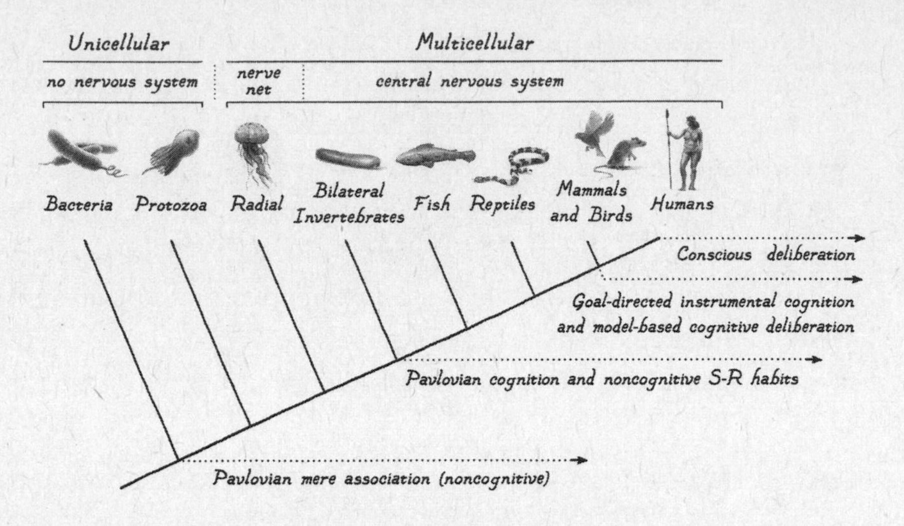


**Figure 43.2:** ***Cognition and Life***

inherited it from a common ancestor? In favor of this possibility is the fact that molecular mechanisms, and even genes, underlying Pavlovian conditioning are highly conserved in protostomes (worms, mollusks, flies) and vertebrates (mice, rats).

A more elaborate kind of cognition occurs in instrumental conditioning. In Pavlovian conditioning the conditioned response is usually an innate response that comes under the control of a new stimulus. But in instrumental conditioning a novel response is acquired as a result of its reinforcing outcome. As we've seen, a common approach involves placing a mildly food-deprived animal in a chamber with a lever. Through random movements they eventually touch the lever, and a food pellet is delivered. Over time they learn to press the lever to get food.

Anthony Dickinson showed that there are two different kinds of instrumental responses—habits and goal-directed actions. These look the same behaviorally, and can only be distinguished by special tests. For example, if a food-deprived rat receives food when it presses a lever in the presence of a flashing light, it will begin to press the lever when this stimulus appears. At that point, the value of the food is manipulated by controlling the animal's degree of food deprivation (food is more valuable to a hungry than a satiated animal) or its history with food (food that recently made the animal sick is less valuable). If the rat presses the lever despite food being devalued, the behavior is likely a habit; if it presses more with food that the animal values more (sucrose versus plain water), the behavior is goal directed. Habits only require mere association between the response and the food reinforcer—the reinforcer stamps in the response, and when a similar situation occurs later, the response is repeated. But goal-directed responses require a memory of the value of the outcome that stamped in the response in the first place (table 43.3).

TABLE 43.3: Dissociating Two Kinds of Instrumental Actions: Habits versus Goal-directed Behaviors

**1. INSTRUMENTAL TRAINING**

Lever press (CR) during flashing light (CS) → food (US)

**2. INSTRUMENTAL TEST**

Light (CS) → lever press (CR)

**3. US DEVALUATION**

Food taste (CS) + nausea (US)

**4. CONDITIONED RESPONSE TEST**

—The lever press response (CR) is a stimulus-response habit if it occurs despite food devaluation, since it does not depend on the current value of the food

—The lever press response (CR) is a goal-directed response if it fails to occur after devaluation since it depends on the current value of the food

ABBREVIATIONS

**CS**—conditioned stimulus

**US**—unconditioned stimulus

**CR**—conditioned response

So which animals have the cognitive form of instrumental learning? Like Pavlovian conditioning, instrumental conditioning is widespread among vertebrates (see figure 43.2). However, there is no convincing evidence that goal-directed instrumental behavior occurs in early vertebrates (reptiles, amphibians, fish). Some claim to have demonstrated goal-directed instrumental learning in invertebrates, but the evidence mostly involves the modification of innate behaviors rather than the learning of a novel behavior on the basis of the value of its consequences. Compelling evidence for goal-directed behavior has really only been found in mammals and birds.

This conclusion does not absolutely rule out the possibility that some form of cognition may one day be found in protostome invertebrates. But even if it exists, it would not likely be relevant to cognition in vertebrates since, so far at least, only mammals, and some birds, have been shown to have the capacity to use internal representations of goals to flexibly respond to changing environmental situations. If this capacity is present in protostomes, but lacking in fish, amphibians, and reptiles, it is unlikely that whatever is found in protostomes is what underlies behavioral flexibility in mammals and birds.

Why do some animals have two kinds of instrumental responses? In other words, why form habits if you have evolved the ability to flexibly respond? Doesn't that just get you into trouble, reverting to a more primitive and rigid kind of responding? In their book *The Evolution of Memory Systems*, Elisabeth Murray, Steven Wise, and Kim Graham proposed that the two behaviors actually complement each other. Habits work well when the environment and its resources are stable. But when the environment is volatile and resource access less predictable, it is very useful to be able to respond flexibly on the basis of opportunities available to satisfy current needs, depending on what worked recently in similar situations.

One of the key ways that animals use internal representations formed through instrumental learning in daily life is in object recognition. Perception of meaningful stimuli requires that present sensory information be interpreted on the basis of templates stored in memory. These memory templates are established by the reinforcing outcomes of stimuli, and are called upon when foraging for food and drink, and avoiding harm.

The habit-versus-action story is instructive in considering other claims about animal cognition. For example, there is a large body of research on so-called higher cognition in invertebrates. Particularly impressive are studies that suggest that invertebrates can form concepts and use them to make behavioral choices. However, Martin Giurfa, the lead author of a key study of conceptual learning in bees, later argued that simpler elemental (mere associative) explanations can account for such behaviors. This conclusion is consistent with other findings showing that in these organisms the circuit underlying the supposedly higher cognitive forms of learning is the same as that underlying simple, noncognitive learning based on mere associations.

We are thus left with the conclusion that while Pavlovian cognition is widespread among animals, the capacity to use internal representations of goals to flexibly respond to changing environmental situations is far more limited, only clearly having been shown in mammals, and to some extent, birds. If correct, this conclusion has profound implications for our understanding of the course of vertebrate brain evolution.

*Chapter 44*

# THE EVOLUTION OF BEHAVIORAL FLEXIBILITY

Goal-directed instrumental learning is sometimes talked about in terms of natural selection at the level of the individual. In other words, trial-and-error learning enhances fitness by selecting adaptive behaviors in a single animal, much like natural selection enhances the fitness of the species through the genetic selection of adaptive body traits.

Behaviors can be learned in novel situations when they have beneficial outcomes, such as obtaining food or drink when energy or fluid supplies are low, or preventing pain or other harm in the face of danger. The fact that the behaviors can be arbitrary (i.e., have no particular relation to the goal) is the key to behavioral flexibility.

The usual explanation for why behaviors that attain such goals are learned is that they have emotional consequences. Darwin said as much when he proposed that the emotional states of mind that occur in survival situations help animals identify actions that are useful in their quest to survive. Thorndike made this a formal learning principle, called the "law of effect," noting that behaviors that lead to satisfying (pleasurable) consequences are repeated, whereas those that have dissatisfying (painful) consequences are not.

Darwin and Thorndike drew upon the ancient idea of hedonism—that a major driving force in behavior is the attainment of pleasure and avoidance of pain. With the law of effect, pleasurable and painful stimuli became rewards and punishers. Because these terms smacked of subjective emotional implications, the behaviorists introduced the term *reinforcers* to describe the objective events that increase the likelihood that behaviors will be repeated in similar situations in the future (table 44.1).

Table 44.1: Hedonic (pleasure) vs. Nonhedonic (reinforcement) Explanations of Instrumental Conditioning, and Dopamine's Role in Each

**HEDONIC HYPOTHESIS**

CS (sight of cake) + US (taste) → ***pleasure*** → repeat eating of cake

**DOPAMINE → PLEASURE**

CS (sight of cake) + US (taste) → dopamine → ***pleasure*** → repeat eating of cake

ABBREVIATIONS

**CS**—conditioned stimulus

**US**—unconditioned stimulus

**CR**—conditioned response

**NONHEDONIC HYPOTHESIS**

CS (sight of cake) + US (taste) → ***reinforcement*** → repeat eating of cake

**DOPAMINE → REINFORCEMENT**

CS (sight of cake) + US (taste) → dopamine → ***reinforcement*** → repeat eating of cake

Darwin realized that emotional evolution would have to be achieved through the effects of natural selection on the brain, and MacLean's triune brain theory, appended to Edinger's model of brain evolution, made this concrete. MacLean argued that the evolution of the limbic system gave mammals the ability to feel emotions, and to learn responses associated with these feelings.

Not long after MacLean introduced the limbic system, James Olds and Peter Milner found that rats would learn arbitrary behaviors, such as pressing a lever, to receive jolts of electricity to certain brain areas, many in regions of the limbic system (amygdala, hypothalamus), or in brain pathways carrying chemicals like dopamine to these areas. Initially, they described their findings as having identified the physiological basis of the behaviorist principle of reinforcement. But Olds soon went in a different direction. With behaviorism on the wane, Olds published "Pleasure Centers of the Brain," an article that transformed reinforcers back into hedonistic rewards based on pleasure. Pleasure again became the explanation of behavioral flexibility. Subsequently, Roy Wise introduced the idea that dopamine underlies pleasure, connecting dopamine to both pleasure and reinforcement (see table 44.1).

Hedonic states, in their pure form, are tied to sensory receptors that detect particular kinds of stimuli. For example, when receptors in the skin detect tissue irritation or damage, we feel pain. Other skin receptors detect stimuli that are experienced as pleasurable, such as light touch on the back, arms, neck, or genitals. Still other receptors in the mouth or nose detect chemicals that are experienced as pleasant or unpleasant tastes or odors.

While pleasure and pain are often treated as emotions, they are actually different. There are no sensory receptors for fear, anger, sadness, joy, or other emotions—the content is determined by the brain. The stimulus suggests but does not define an emotional experience.

It is important to note that the conscious feeling of pain or pleasure that humans experience when certain receptors are active is but one of many consequences of the sensory signals that reach the brain. They also elicit reflexes or other innate reactions, and increase brain arousal, motivate instrumental behaviors, and reinforce learning. Each of these consequences, including the conscious feelings, has separate neural underpinnings, and we should not assume that observation of one of the nonconscious behavioral consequences means that a conscious feeling of pain or pleasure has occurred. In animals, all we can measure are the behavioral consequences. Measuring the conscious manifestations is far more difficult, a subject we will discuss thoroughly later.

That we should be cautious in how we consider hedonic states is suggested by the fact that if a person with chronic pain is distracted by a funny joke, he does not experience the pain while laughing. The nociceptors are still responding, but the subjective pain is not noticed. (Distraction also helps explain why hypnosis can be useful to such patients.) Additional caution comes from studies of addictive drugs, which are now known to lead to compulsive use because they hijack habit circuits in the brain, not because of the pleasure that humans can experience when they initially use the substance.

The idea that hedonic feelings are what make flexible instrumental learning possible matches our everyday experience—we feel good when rewarded and bad when punished. But the question is, Are these feelings the actual explanation for why behaviors that attain goals like food or avoidance of pain are learned and repeated? I think there's a different explanation, one that does not depend on hedonic feelings.

Behaviorally, reinforcers are stimuli that change the value of some other stimulus (in Pavlovian conditioning) or of a response (in instrumental conditioning). For example, during Pavlovian conditioning, the reinforcing unconditioned stimulus alters the ability of the conditioned stimulus to activate neurons to which it is synaptically connected. In instrumental conditioning, the reinforcing unconditioned stimulus establishes a relation between neurons that process a stimulus and a response, making it more likely that in the presence of the stimulus the response will occur. If the responses depend on the unconditioned stimulus being a valuable outcome at the moment, then the response is goal directed; otherwise, it is a habit. Chemicals like dopamine regulate these processes and play a key role in reinforcement learning through their neurobiological actions on neurons, but not because pleasure, pain, or other hedonic states intervene. Just as Olds came to question the pleasure center idea, Wise retracted the dopamine pleasure principle.

Given that mammals and birds are the only two groups that clearly exhibit flexible goal-directed learning, and they descended from different reptilian ancestors, it is likely that these groups developed this capacity independently. Let's try to piece together the history.

Early vertebrates came equipped with key behavioral tools. They could form mere associations between meaningless and meaningful stimuli, and could also use stored internal (cognitive) representations to learn more complex kinds of Pavlovian associations. They could even learn instrumental responses that produced useful outcomes, but only in a rigid, habitual way. They were unable to take the step that mammals and birds took toward flexible learning of novel responses acquired by storing a representation of the value of the outcome. While habits are often acquired when goal-directed actions are repeated many times, they can also be learned in mammals without going through a goal-directed phase. In early vertebrates this direct route of habit learning was likely the only path to instrumental behavior.

I think that the usual explanation of how flexible instrumental behavior evolved—that the evolution of the limbic system gave mammals emotional feelings that they could use to assess the good and bad situations in the world—is wrong, and I have a different hypothesis.*

With the extinction of the dinosaurs, mammals were able to forage freely and to explore new niches with less risk from predation. This more nomadic lifestyle posed challenges that placed novel pressures on the brain. The result was the fusion of the ability to use cognitive representations in Pavlovian conditioning with instrumental learning abilities. What's the evidence?

We know that present-day reptiles possess the ability to store internal representations of toxic foods for use in the future via Pavlovian conditioning. If this was also present in the reptilian ancestors of mammals, it could have been a foundation, under the right kinds of selective pressures, for the emergence of goal-directed behavior in mammals, by transforming rigid stimulus-response habit learning into learning based on behavioral outcomes. This would have enabled mammals to retain memories of the success and failure of past foraging episodes, including information about

* Keep in mind that a lot of guesswork goes on when one talks about why something happened in the evolutionary past.

where food was found previously, information about the respective value of food items acquired in different locations, and the efficiency and risks of different routes.

Peter Dayan has pointed out that it is remarkably simple to account for how behavior is learned by its consequences. He draws on findings from his field, called reinforcement learning, which explores how artificial systems learn to optimize their behavior on the basis of outcomes. Learning by consequence only requires that the learning agent has a way to create a representation of its last action, its state when it chose that action, and the payoff received (the value of the outcome). Value in this context is not an emotion or feeling, but simply a quantitative representation of the payoff.

In artificial agents the computations are performed by equations. Animals use cells, synapses, circuits, and molecules in their nervous systems to do this. A key player in these neural computations of value is dopamine, which, when released onto neurons forming associations between stimuli and between stimuli and responses, strengthens the connections. This effect of dopamine in computing value has nothing to do with pleasure, but rather with tweaking the cellular responses of neurons. Pleasure, when it occurs, is a correlate, not primarily a cause, of learning.

In some organisms, certainly humans, pleasure can contribute to the reinforcement of learning. But we cannot generalize from our inner experiences to account for behavior in organisms whose inner experiences are not knowable by us, especially when nonconscious explanations suffice.

Kent Berridge and Morten Kringelbach, leading hedonic researchers, argue that the neural underpinnings of behavioral activities commonly used to identify hedonic states in animals reflect mechanisms that are so basic and so essential to survival that they may well have evolved long before the additional mechanisms that make conscious feelings possible in humans. In other words, hedonic states that control behavior are fundamentally nonconscious states. As we will see later, such nonconscious states, can, in some organisms, be re-represented by cognitive systems in such a way as to be consciously experienced. But I don't think that such conscious feelings are what enabled behavioral flexibility.

Going against conventional wisdom, I propose that the evolution of a new cognitive capacity—the use of internal representations in conjunction with instrumental learning—was what made behavioral flexibility possible. This capacity for deliberation allowed the stored value of goals from the past to be used to guide present behavior in novel and powerful ways.

PART 10

# Surviving (and Thriving) by Thinking

*Chapter 45*

# DELIBERATION

The ability to use internal representations in conjunction with instrumental learning was a significant change in the behavioral tool kit of animals. It endowed its possessor with a novel talent—the capacity to base current responses on memories about the past consequences of successful trial-and-error learning. But suppose an animal with this ability finds itself in a situation that is occurring for the first time in its life. The best option it has is to start a new round of trial-and-error learning. But when life or well-being is on the line, trying out random, untested options on the chance that one will pay off is likely to produce failures, with possibly fatal consequences. At some point in the evolutionary history of mammals, the capacity for inner deliberation emerged* (table 45.1).

TABLE 45.1: How Is Goal-Directed Deliberation Different from Instrumental Learning?

| GOAL-DIRECTED INSTRUMENTAL LEARNING |
|---|
| —Retrospective: outcomes of past trial-and-error learning inform present behavior |
| —inefficient in novel situations |
| **GOAL-DIRECTED DELIBERATION** |
| —Prospective: present behavior based on predictions of future outcomes |
| —allows creative solutions to novel problems |

* This chapter is based in part on a paper I coauthored with Nathaniel Daw, "Surviving Threats: Neural Circuit and Computational Implications of a New Taxonomy of Defensive Behavior."

Deliberation allows one to envision possible response options and to use practical knowledge stored in memory to evaluate and choose the one that seems most likely to produce a useful outcome. Additionally, through deliberation, disparate pieces of information can be combined across multiple steps of reasoning to rapidly reach a decision. What would take many repetitions of trial-and-error behavior for a new response to be learned via outcome-based reinforcement can be compressed into a moment of internal simulation of the potential outcomes of possible choices.

Consider an example. Suppose you find yourself trapped by a bear on a riverbank. Having never been in this situation before you do not have an instrumental response to draw upon. However, you can use practical reasoning based on general knowledge about the kind of situation you are in to generate novel hypotheses about what might work best. You might remember having been successful in the past in escaping from other kinds of danger by running away, and generalize from that experience to the present one. But you might also consider that you've heard that bears are really fast runners and conclude that running might not be a good option. You then weigh the choices of swimming or climbing, and assess your talents at these options versus the bear's, and choose the one you conclude has the better chance of resulting in a desirable outcome.

Just as the determination of whether a behavior is a goal-directed action depends on ruling out that it is a stimulus-response habit, the demonstration that a behavior depends on deliberative cognition requires that it cannot be explained in terms of either the habitual or goal-directed forms of instrumental responding. Unlike processes that underlie habits, deliberative cognition is effortful, as opposed to automatic. Deliberative actions are related to outcome-dependent actions, in that they are flexible and goal-oriented. But responses chosen through deliberation are not based on a historical relation between the individual having experienced firsthand a particular response and its physical outcome. Instead, deliberative cognition mentally simulates possible future outcomes when planning and strategizing about how to accomplish a goal—in such cases, in fact, previously learned actions and habits may have to be suppressed if a novel action is to be taken.

Deliberation is said to utilize mental models. Perhaps the best-studied example of a mental model is a spatial map acquired by accumulating knowledge about the relationship between particular landmarks. Many animals, including bees, birds, and mammals, use such maps when foraging for food, avoiding danger, or simply getting around. But in deliberate cognition, these maps are not simply passively followed. They can also be used to generate and compare options. For example, with deliberation a route can be planned by mentally simulating both what might be the most efficient and the safest choice. Within mammals, primates have deliberative facilities that other mammals lack. Humans take it to another level altogether, in part due to the benefits that accrued from having evolved a more sophisticated brain, including the cognitive boost to the human brain conferred by language.

*Chapter 46*

# THE ENGINE OF DELIBERATIVE COGNITION

Deliberative cognitive processes in humans are often described as being dependent on what Alan Baddeley termed working memory (figure 46.1). In contrast to long-term memory, working memory involves the transient storage of task-relevant representations for use in mental work. It is through working memory, and its so-called executive control functions, that information can be held in mind temporarily in the process of deliberating.

Executive functions allocate and focus attention on selected sensory stimuli in the external environment or from within one's body; they re-

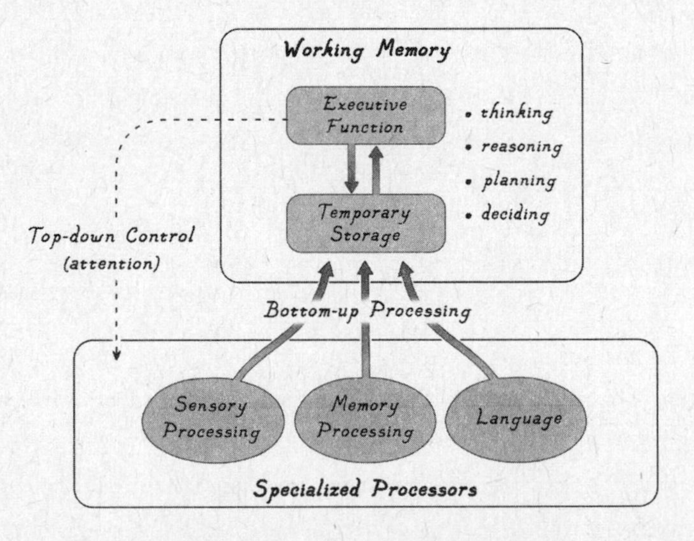


**Figure 46.1:** ***Working Memory***

trieve relevant memories; and they orchestrate the maintenance of the selected sensory and mnemonic information in a temporary active state so that it can be evaluated and integrated with other information to form new representations and use them in thinking, reasoning, planning, and deciding.

Executive functions are prospective—they make deliberative predictions about possible future states of the world in the process of achieving goals. But the cognitive activities used to achieve goals often involve many steps, and by necessity extend over time. This is done by selecting and integrating information, dynamically updating what is being processed depending on progress toward the goal, and making adjustments when unforeseen obstacles occur.

Two key features of working memory are thus its ability to temporarily maintain representations and to integrate diverse kinds of representations into new unified representations. The temporary maintenance of information in working memory enables you to remember someone's name when you are introduced, and integrate her name with her appearance and voice to create a unified representation, a model of that person that allows you to recognize her later by sight or sound, and to recall her name.

Another important feature of working memory is that it has a limited capacity. Most people can hold in mind four to seven separate items. As the number increases, working memory capacity decreases. Remembering the ten items that constitute a U.S. phone number is difficult, especially if you are given them in the course of a conversation where other demands are being placed on working memory.

A newer view is that working memory is not limited by the number of items per se, but by the amount of information being distributed among the items. In this position, the quality rather than the quantity of representations underlies the capacity limit of working memory.

One way to overcome the capacity limit is to bundle information into meaningful chunks. For example, you can divide a phone number into three groups (area code, local exchange, number), each consisting of three or four items. If you are familiar with the area code you can concentrate on the other numbers, simplifying the task. The reason you can off-load a familiar area code is because it is part of a stored body of knowledge.

Bundles of stored information (memories) about related items are called schema. This term was first formally introduced in psychology by Sir Frederic Bartlett to account for how existing knowledge both influences the acquisition of new, unfamiliar information and affects how it is remembered. Later, the developmental psychologist Jean Piaget adopted the notion of schema to help account for how, mental step-by-step, a young child assembles an understanding of the world and develops competence in dealing with it. He proposed that as the child accumulates related experiences in memory, the schema is updated and becomes more elaborate, a process called assimilation. If subsequent experience contradicts the theme of a schema, the schema will often change, or the contradictory event will be reinterpreted to be consistent with the schema; this is called schema accommodation (figure 46.2). John Bowlby, another influential developmental psychologist, also emphasized schema in the buildup about knowledge, especially about emotions and mother-infant attachment.

Schema are closely related to the idea of mental models introduced in

**Figure 46.2:** ***Schema Updating by Assimilation and Accommodation***

the last chapter. Their entry into cognitive science was facilitated by the work of the artificial-intelligence pioneer Marvin Minsky, who suggested that schema (or frames) could be used in computer models to simulate top-down cognitive effects used by humans in information processing. Indeed, people use schema to generate expectations about a given situation. They facilitate thought, reasoning, and decision making, enabling the assessment of opportunities and risks of different options.

Many kinds of schema are formed and used in daily life. Some concern features of stimuli in the world (the color or shape of visual objects like apples or hammers, and what actions go with apples or hammers). Still others are more conceptual and involve situations (work, party, vacation), people (parent, employee, Democrat), psychological states (emotions, thoughts, beliefs, desires), and one's self (views of one's self and one's typical ways of thinking about things or ways of acting in certain situations). Stereotypes (about age, gender, race, ethnicity) are also mental models/schema that influence how you interact with people of a certain background. In the context of psychopathology, Aaron Beck, the father of cognitive theory, proposed that in mental disorders, such as depression, one develops a mental model or schema of being a sad or excessively worried person, and becomes focused on negative self-thoughts. Jeffrey Young and Robert Leahy have introduced schema-based approaches to cognitive therapy.

Schemas have been called the building blocks of cognition. They allow new information to be conceptualized based on past experience, and to be stored and remembered more efficiently. Executive control functions of working memory play a key role in the categorization, updating, and storage process, and also in the later retrieval of the information in relevant situations in the future (figure 46.3).

Schema work their magic by taking advantage of the ability of the brain to complete patterns from partial information, a process called pattern completion. For example, the opening phrase of a piece of music is sufficient to bring to mind the overall sound of the song, the genre of music, the artist, where you were when you heard it first, and so on. And

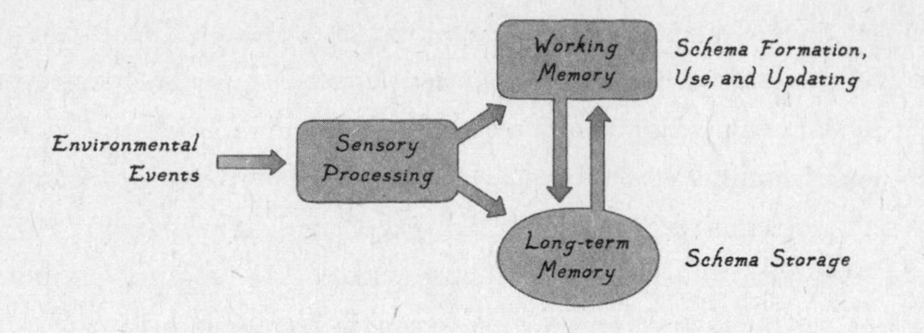


**Figure 46.3:** ***Working Memory Updates Schema Stored in Long-Term Memory***

if you notice a small part of a roundish, orange-colored item, you easily recognize it as the fruit called an orange, especially if you are in the fruit section of a grocery store. Not surprisingly, brain circuits involved in working memory are implicated in pattern completion of schema.

Pattern separation is complementary to completion—to know what something is, one also needs to know what it is not. Cats and dogs are both furry animals that walk on four legs but are different from other classes of mammals, like rats or bats; pit bulls and poodles are both dogs, but very unlike in appearance and temperament. The capacity of the human brain for cognitive pattern processing is unique in the animal kingdom. As noted by Patricia Alexander, "Without the ability to discern meaningful patterns in the stream of data that continually flood the senses, humans would remain prisoners within a world of isolated sights, smells, and sounds, unable to comprehend or to build on experiences across time and space."

More so than other animals, humans have the ability to use working memory to integrate information across sensory modalities, to think conceptually and schematically, rather than just perceptually, and to orient toward the future. We experience our lives as unified events rather than as a collection of sights, sounds, and other isolated perceptions.

The ability to form concepts, though hardly unique to humans, has unique qualities in our species. For example, Derek Penn, Keith Holyoak, and Daniel Povinelli argue that an important distinctive feature of human cognition is the ability to use concepts to engage in hierarchical relational reasoning—humans can easily understand why the relation of "bird" to "nest" is the same as that of "dog" to "doghouse." Relational reasoning, also called cognitive branching, makes it possible for us to simultaneously consider multiple options—to multitask—when engaging in reasoning and problem solving.

One feature of human cognition deserves special consideration. This feature, more than anything else, distinguishes human capacities from the deliberative abilities of all other animals—and that is language.

*Chapter 47*

# SCHMOOZING

Do you talk to yourself? Most people do. We use a kind of inner monologue, sometimes referred to as "self-talk," when thinking, daydreaming, making decisions, solving problems, or regulating our emotions, and when weaving narratives to account for behavior controlled nonconsciously. And in social situations we can simulate a dialogue in which we take the point of view of others (a kind of mental model about other minds). Writers do this quite often, envisioning how their words will sound when read by either some mythical all-purpose reader, or some specific kind of reader.

The idea of inner speech was made famous by the Russian psychologist Lev Vygotsky. He noted that it is not quite the same thing as ordinary spoken language, as it is not as formal or rigid. Vygotsky was interested in how children acquire and use inner speech in the process of cognitive development. As explained by Oliver Sacks in *Seeing Voices,* "It is through inner speech that the child develops his own concepts and meanings; it is through inner speech that he achieves his own identity; it is through inner speech, finally, that he constructs his own world." Language and deliberative thought, and even consciousness, are closely entwined.

Aldous Huxley noted that it is via language that "we have raised ourselves above the brutes." Indeed, many in the psychological and brain sciences agree that the human capacity for language is key to the complex, rich, and unique nature of human cognitive mental life. But others insist that language can't be the answer. A common argument used against the language and cognition connection is that deaf people and people who lose speech because of brain damage aren't unconscious zombies. It's not the ability to talk per se that is key. What matters is what underlies talking—what language does for cognition. In *Kinds of Minds*, the philosopher Daniel Dennett put it this way: "The kind of mind you get when you add language to it is so different from the kind of mind you can have without language that calling them both minds is a mistake." Elsewhere, Dennett says that "language lays down tracks on which thoughts can travel."

Recall, the Greeks sought to carve nature at its joints by classifying the natural world. They could do this because they had language. Much later, Benjamin Whorf wrote: "We dissect nature along lines laid down by our native languages. . . . Observers are not led by the same physical evidence to the same picture of the universe, unless their linguistic backgrounds are similar." Whorf is partly responsible for the most famous idea about the relation of language to thought. The Whorf-Sapir hypothesis, for example, emphasized the role of language in shaping perceptual experience. This idea fell out of favor under the scrutiny of Noam Chomsky, the powerful and opinionated linguist, and his influential student, Steven Pinker. Jerry Fodor, who provided a philosophical foundation for cognitive science in its early days, also rejected the idea that natural language is the language of thought, and instead introduced the idea of "mentalese," a kind of nonconscious universal language in which we do our thinking. But with some modifications made in light of new findings, Whorf's notion that language and culture shape thought and experience is currently thriving again in psychology.

Language allows thoughts to wander in novel directions and yet stay connected as a "train." It provides words to label external objects and to characterize and recognize our perceptions, memories, concepts, thoughts, beliefs, desires, and feelings. The words individuals use reflect the things of significance in their culture. For example, Whorf made famous the notion that people living in snowy environs have names for and can recognize more kinds of snow than those not living under such conditions, because snow is important for their ability to survive and thrive.

But language does much more than simply name and categorize objects and events and organize their underlying conceptions. With language also comes syntax, or grammar, which structures our mental processes and guides their operation when we are thinking, planning, and deciding. The cognitive neuroscientist Edmund Rolls has noted that syntax enables humans to plan actions and evaluate their consequences by anticipating many steps ahead, without having actually to perform the actions. (This is a version of hierarchical reasoning.) Most other animals, Rolls notes, are limited to innate programs, habits, and rules, or, in the case of mammals and birds, to reinforcement-based instrumental learning. We can give primates a bit more cognitive credit because of their greater facility with deliberation to solve problems. But without the ability to bring language into deliberation, thought remains static and crude.

Many animals can even communicate with one another about specific things. Birds use vocalizations to attract mates, identify their young in group-living situations, and recruit colleagues to outnumber ("mob") predators. Monkeys and apes take the process a step further, using different vocal expressions to identify different predators (cats versus hawks, for example). But if by language we mean the capacity to flexibly use sounds or visual symbols to spontaneously indicate things about the present, past, and/or future, then only humans possess that ability. Humans alone can call upon syntax to convey to other humans exactly *when* and *where* a particular kind of predator (*what*) was seen today, and then discuss the implications with the others for the purpose of making a plan about how to act tomorrow.

Peter Godfrey-Smith and others minimize the importance of language to cognition, arguing that complicated mental processes occur in animals without speech. Indeed, animals do have sophisticated cognitive capacities, but no other animal matches humans in abstract conceptual thought, hierarchical relational reasoning, and pattern processing. Paraphrasing Dennett, language is not necessary for cognition, but cognition without language is not the same as cognition with it.

Mark Mattson has described language as the quintessential example of the superior capacity of the human brain for pattern processing: "Language involves the use of patterns (symbols, words, and sounds) to code for objects and events encountered either via direct experience or communication from other individuals." Language, he says, "can create new patterns (stories, paintings, songs, etc.) of 'things' that may (reality) or may not (fiction) exist."

Language also greatly facilitates our ability to think conceptually and schematically, rather than just perceptually. For example, Marilyn Shatz, an expert in early language acquisition, notes that "Animals, clever though they are, remain mired in re-description of perceptually based data. . . . Much of the immense power of language stems from the ability of the language user to engage in conversations with other like-language users to obtain more material on which to exercise one's higher-level cognitive abilities."

Without language our capacity for hierarchical relational reasoning, the ability to reason across categories, would be greatly diminished. Together language and relational reasoning revamped the nature of cognition—humans can think about linguistic concepts like "I," "me," and "mine" in relation to the future and past, and to others. Social communication, shared values, coöperation, and culture were the result. The linguist Dan Everett calls language "the cultural tool."

Michael Corballis, for example, proposed that language evolved because it allowed communication about internal states to others. This appealing idea, if true, would have made it possible for one person to tell another about past experiences, saving the listener the need to learn the same information about foods that were particularly tasty or that caused illness, or about the success of certain behaviors in capturing prey or avoiding predators. Similarly, if one person was able to figure something out by deliberating (reasoning) about a problem using hierarchical mental simulation, such as how to build a novel kind of shelter, this knowledge could be passed on. These kinds of exchanges could well have been the foundations of the schema and concepts that constitute folk wisdom and culture. And without language, the preservation of such wisdom and culture across generations would not be possible.

Robin Dunbar has proposed a related idea about the early social benefits of language—namely, that it made it possible to share not only knowledge about foods and foes, but also about other humans: to gossip. With language, our early human ancestors could exchange information about who was and wasn't trustworthy, or what traits made a good mate. He expands on this, noting that most human communication today is gossip. When people interact socially, he notes, they often get around to gossip: who in the community is a jerk and who's not, who's cheating on his or her spouse, and so on.

A strong position is also taken by Evan MacLean on the role of social factors in human cognitive evolution. He argues that what sets human cognition apart is our ability to transcend competitive impulses and engage in cooperation, to reason about the intentions and desires of others of our kind, and to communicate with one another linguistically. This capacity, he says, is not hardwired by genes, but depends on the accretion of incremental learning across generations; in other words, culture.

Consistent with this line of thought, Michael Tomasello and Hannes Rakoczy have argued that a human child, living alone from birth on an island, lacking the benefit of the incremental learning that constitutes cultural history, would have a cognitive profile more like that of a sophisticated chimpanzee than an adult human—human genes, in other words, are not alone sufficient to make a human mind. Also needed is the verbal history of the culture.

Yuval Noah Harari proposed that by changing the nature of human cognition, language made culture possible, especially by enabling representations of not just what is present but also what is absent. One kind of absence particularly relevant to culture is the memory of people from the past, be they family members or friends, or impersonal cultural icons. (Think of the impact of the death of Princess Diana on the United Kingdom.) Language also makes possible the creation of cultural myths (e.g., the invincibility of a leader; the lion as a protector of the human spirit; deities and religions).

Cecilia Heyes has recently proposed that what separates humans from apes is our "cognitive gadgets," such as "mind reading," imitation, and language. These, she argues, are specialized psychological innovations that emerge, and are maintained across generations, not by genetics, but by social learning and culture. A related idea proposed by Nick Shea and colleagues is that culture was, in part, made possible by the unique ability of our brain to represent its cognitive processes and share them with others.

I agree that culture has played a crucial role in separating humans from the great apes and making possible what humans have achieved throughout history. But I don't think we should rule out a role for genes and natural selection in this process.

Some years ago, Stephen Jay Gould pointed out that some useful traits emerge as by-products of adaptive (naturally selected) traits. He called these *exaptations*. Once such a by-product emerges, if it ends up being useful, it can increase in frequency among those who survive and reproduce and become part of the species endowment via natural selection. Feathers, for example, arose in terrestrial reptiles as a source of warmth, but then were beneficial for flight, and are clearly maintained via natural selection in present-day birds for this purpose. Oren Kolodny and Shimon Edelman recently proposed that language arose this way. Specifically, they suggest that language was cobbled together in early humans by way of synaptic plasticity that coupled together neural mechanisms underlying existing traits, such as nonverbal communication, serial cognition, and tool use.

Kolodny and Edelman's intriguing idea helps bridge the divide between biological and cultural explanations of the origin of language. For example, once the systems were coupled, many of the souped-up features of human cognition discussed in this chapter could have resulted in enhanced pattern processing, conceptual thinking, and hierarchical deliberation; communication about internal states; mind reading; gossip; and culture. With the ball set rolling, adaptive synergy created by the various cognitive capacities related to language likely drove, and continues to drive, additional changes in the brain via synaptic plasticity in the short-term and via natural selection in the long-term.

The philosopher Ludwig Wittgenstein famously said that if a lion could speak, we wouldn't understand what he was saying. Wittgenstein probably meant we wouldn't understand the lion because its frame of reference would be completely different from our own. Certainly the absence of cultural concepts and schema would be an impediment. But the key question for us here cuts deeper: it is whether a linguistically competent lion would be cognitively similar to humans, simply because of its speech. Probably not. To have the cognitive capacities we possess would require its having a brain that went through the specific set of neural adaptations under the specific selective pressures that the brains of our ancestors did, including the pressures provided by human culture. A talking lion would still be a lion.

## Part 11

# Cognitive Hardware

*Chapter 48*

# PERCEPTION AND MEMORY SHARE CIRCUITRY

Human cognitive deliberation, mental models, schemas, pattern processing, conceptualization, hierarchical relational reasoning, language, and the like are products of the kind of brain we have. And within the brain, cognition is to a large degree a product of our cerebral cortex, especially the laterally located neocortex.

Anatomists have traditionally divided the neocortex into four main lobes: occipital, temporal, parietal, and frontal (figure 48.1). But in order to understand the neural basis of cognition, we can't stop at such a coarse level.

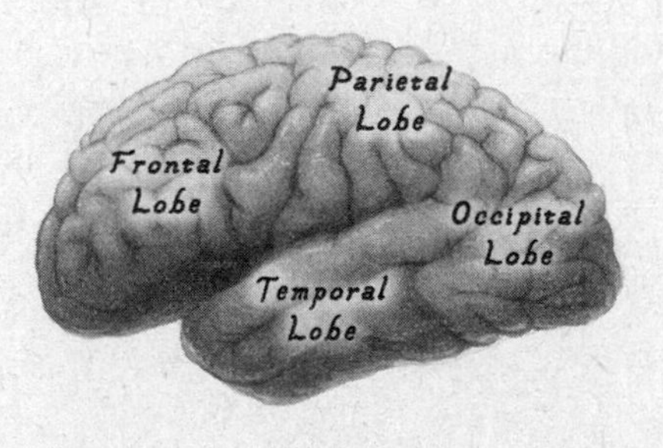


**Figure 48.1:** ***Four Lobes of the Human Brain***

Much of our understanding of cognition is based on how we use sensory information in relaton to thoughts, decisions, and actions. An excellent way to begin to explore the cognitive brain is thus by considering how sensory information received by receptors on the body surface (eyes, ears, skin, nose, and mouth) is processed by cortical areas. I will focus on the visual system as an example, since so much is known about it, but similar principles apply to other sensory modalities.

Visual processing begins when information about the visual world is detected by receptors in the retinae of eyes, and then transmitted through the subcortical visual pathways to the primary visual cortex, located in the occipital lobe.

Within the primary visual cortex there are distinct circuits for processing different incoming elemental features (shape, color, depth, texture, motion, etc.). These low-level visual circuits transmit to secondary areas of the visual cortex that form two distinct visual processing streams* (figure 48.2). The object processing stream (also called the ventral stream) extends from the primary visual cortex into secondary areas in the occipital lobe and on to additional secondary areas in the temporal lobe, especially the inferior (ventral) temporal lobe. This stream consists of a hierarchical sequence of circuits that integrate low-level features about shape and color into higher and higher levels of feature binding that ultimately create synthetic high-definition representations of recognizable objects. The second visual pathway is the spatial-action visual stream (also called the dorsal stream). It extends from the primary visual cortex into secondary areas in the occipital and parietal lobe, and uses depth and motion information to represent the location of the object in space, enabling body movements

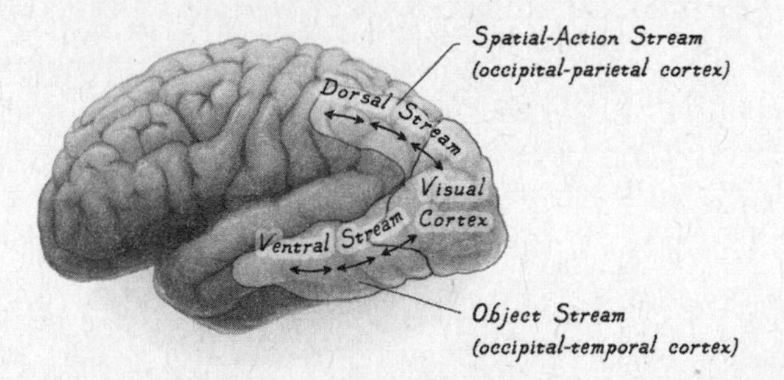


**Figure 48.2:** ***Two Visual Processing Streams***

---

* Some conceptions recognize additional distinctions, such as tertiary areas, but for our purposes we classify all areas beyond the primary as secondary.

to be directed to the object. Each area in each stream connects back to the preceding lower areas, creating processing loops that actively build and modify sensory representations in real time.

The identities and meanings of most objects in the world are not innately wired into the human nervous system. To recognize an object, its features have to be matched with stored templates that have been acquired via experiences with objects of that type. Therefore, in the process of building up perceptual representations of the objects that we "see" and interact with in our lives, the visual system partners with memory systems.

The traditional view is that memories are acquired and stored by the so-called medial temporal lobe memory system, which consists of several cortical areas. These are not in the laterally located neocortex but in medial paleocortical areas—the neocortex is like the outer brown toasted part of a hot-dog bun, and the paleocortex like the white untoasted part that can only be seen by pulling open the two halves of the bun.

The medial paleocortex, you'll recall, was said by Paul MacLean to be the cortical foundation of the limbic system, and home to emotions. One important mark against this idea was the discovery of the central role of paleocortical areas of the medial temporal lobe, especially the hippocampus, in memory. Key paleocortical areas involved in the medial temporal lobe memory system, besides the hippocampus, are the perirhinal, parahippocampal, and entorhinal regions (figure 48.3). These areas contribute, in different degrees, to semantic memory, which underlies object

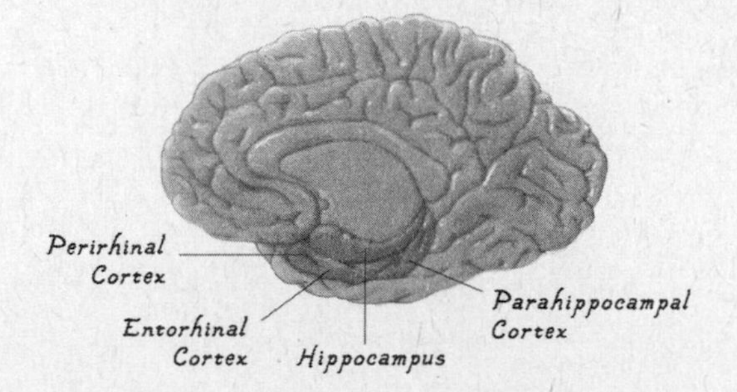


**Figure 48.3:** ***Medial Temporal Lobe Memory System***

recognition, and episodic memory, which underlies the memory of complex events, or episodes, in our lives. (We will discuss these forms of memory in detail later in chapter 57.)

The medial temporal lobe memory areas are intimately connected with the two visual streams (figure 48.4). The object recognition stream continues from secondary areas in the neocortical temporal lobe to the perirhinal cortex, and from there, to a specific part of the hippocampus. This combination of perceptual and memory circuits contributes to semantic memory. The spatial-action stream continues from secondary areas in the neocortical parietal lobe to the parahippocampal cortex, and from there, to different parts of the hippocampus. This combination of perception and memory circuits contributes to episodic memory. The two streams then converge in still another region of the hippocampus, wherein memories are formed about objects in relation to the spatial scene or context in which they occurred.

This distinction between perception and memory circuits reflects the long-standing view that sensory processing ends in the secondary areas of

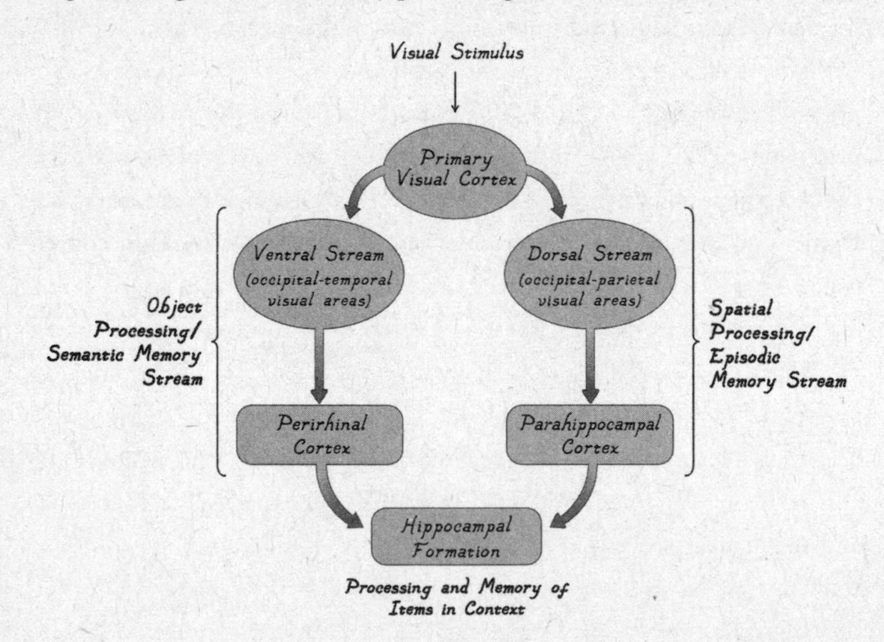


**Figure 48.4:** ***Information Flow from Sensation to Memory***

the visual cortex and that memory processing takes over in the medial temporal lobe. But leading researchers have begun to argue that this view is no longer tenable. Elisabeth Murray, Steven Wise, and Kim Graham reviewed the evidence and concluded that secondary sensory areas are not simply perceptual processors. They store memories about object categories, like faces, tools, or buildings. Further, areas of the medial temporal lobe are not just involved in memory but also in perception (for example, the perirhinal cortex contributes to object perception and the parahippocampal cortex to scene perception). Rather than thinking of these as separate perception and memory systems, many in the field are now thinking about integrated perception/memory streams.

Further emphasizing the relation of perception and memory is the fact that semantic memory is no longer thought to simply be stored in the perirhinal and hippocampal areas of the medial temporal lobe. Areas of the neocortex also store semantic memories, especially multimodal-semantic memories, which can be thought of as conceptual memories. Like the medial temporal lobe, these multimodal conceptual processing areas are intimately interconnected with unimodal sensory areas.

We can recognize the thing we know as an apple by its appearance or taste, and even its smell or the way it feels in our hand, because we have a broad concept of "apple" stored in our brain. Such concepts are formed by the accumulation of experiences with apples through different sensory modalities and the fusion of the features across sensory modalities in convergence zones that receive inputs from different unimodal areas, especially late secondary sensory areas that combine unimodal features within their given modality. Key convergence zones include the areas surrounding the superior temporal sulcus, parietal temporal junction, and temporal pole (figure 48.5). These zones form abstract conceptual memories, including aspects of schema, and make available multimodal conceptual representations that enhance and supplement processing by unimodal areas. The temporal pole is especially interesting, as it constitutes a conceptual hub that integrates information from the other multimodal areas. Another important multimodal area is the prefrontal cortex, which we will discuss in the next two chapters.

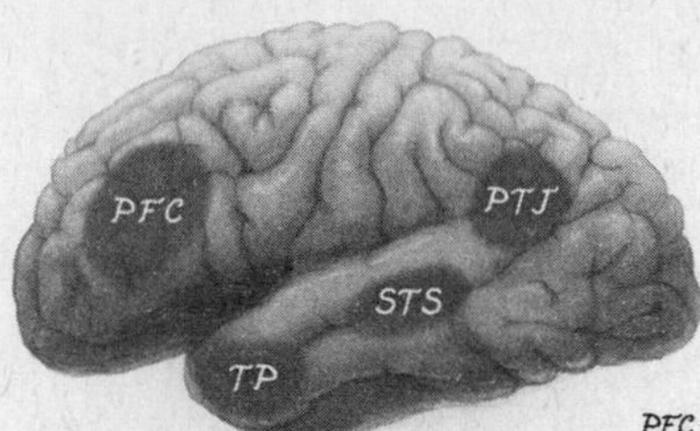


PFC - prefrontal cortex
PTJ - parietal temporal junction
STS - superior temporal sulcus
TP - temporal pole

**Figure 48.5:** ***Multimodal Cortical Convergence Zones***

Language also involves multimodal areas, typically in the left hemisphere. The two areas most associated with language are Wernicke's area, in the parietal temporal junction, and Broca's area, in the frontal lobe. Damage to these results in aphasic disorders in which speech comprehension and/or speech production is impaired. While these areas are essential for communication, the use of language in conceptual processing and relational reasoning also depends on working memory functions of the prefrontal cortex.

The prefrontal cortex is traditionally viewed as home to the headquarters of cognition. This is undoubtedly the case. But it would be wrong to assume that cognition is an exclusive function of the prefrontal cortex. As we see next, prefrontal areas work closely with the posterior cortical areas. Thus, not only are posterior perceptual, mnemonic, and conceptual circuits integrated with one another, they are also closely integrated with prefrontal areas in the service of deliberative cognition.

*Chapter 49*

# THE COGNITIVE COALITION

The prefrontal cortex is the highest level of cortical processing, a hotbed of multimodal integration. The various unimodal and multimodal areas involved in perceptual processing, memory, conception, and language all connect with the prefrontal cortex, which is a major locus of information convergence in the brain, a super convergence zone (figure 49.1). It is, as noted, generally viewed as the central headquarters of working memory—that is, of deliberative cognition. Because the prefrontal cortex receives inputs from the more posterior-located unimodal and multimodal areas, its executive functions can control behavior on the basis of sensory-specific information, stored memories, or highly abstract conceptual representations. And by sending a signal back to these unimodal and multimodal areas, prefrontal circuits can control what these areas process.

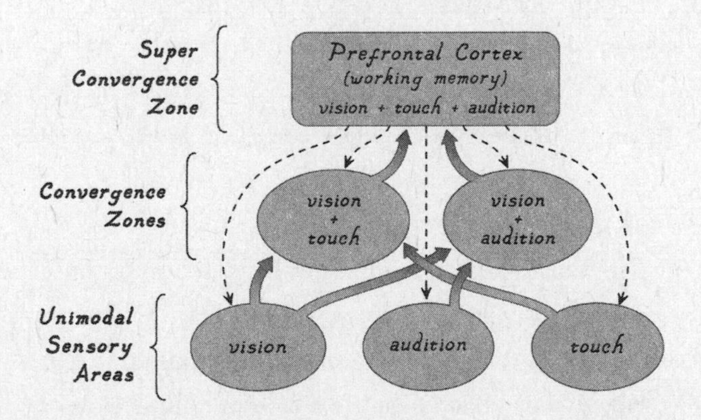


**Figure 49.1:** ***Bottom-Up Convergence and Top-Down Control of Sensory Processing***

Traditionally, both executive and temporary information-maintenance functions of working memory were thought to be products of the prefrontal cortex (figure 49.2). It is now thought that much of the maintenance of information in an active state is done by areas outside of the prefrontal cortex, albeit under top-down control by the executive circuits in the prefrontal cortex (and possibly other areas, such as the parietal cortex). Given this, Bradley Postle, Mark D'Esposito, and Clay Curtis argue that information-maintenance functions of working memory emerge from coalitions of capacities distributed throughout posterior and prefrontal areas of the cortex that are involved in sensory, memory, language, spatial, and motoric processing (figure 49.3). In this way, executive control guides the construction of temporary unimodal and multimodal representations, and maintains them in an active state that can be used in thinking, reasoning, and behavioral control. Different subsets of circuits are recruited, via prefrontal executive control, to solve problems on a case-by-case basis, depending on the task at hand.

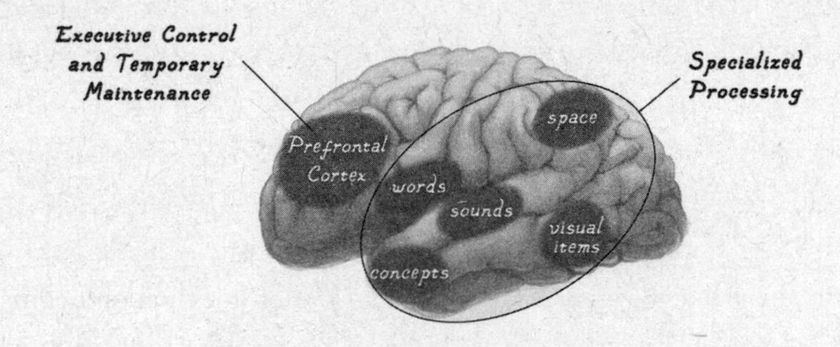


**Figure 49.2:** ***Traditional View of Working Memory in the Cortex***

While much of temporary maintenance is thus in posterior areas rather than the prefrontal cortex in the collation view, the prefrontal cortex does have some role in maintenance. For example, Earl Miller and Jonathan Cohen have proposed that the prefrontal cortex maintains representations of goals in an active state, and uses these to control processing in posterior areas. As explained by D'Esposito and Postle, if you are looking for a friend in a crowd, representation of this goal in the prefrontal cortex biases

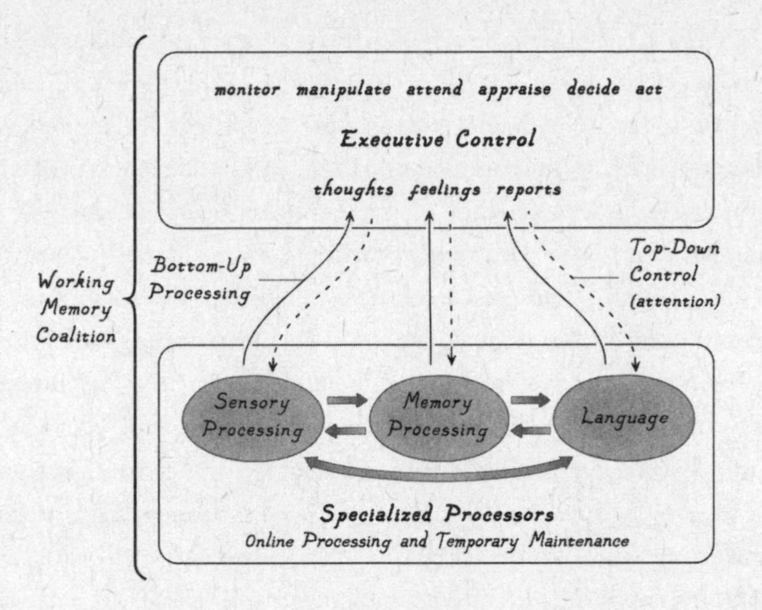


**Figure 49.3:** ***The Working Memory Coalition Hypothesis***

processing in the visual cortex toward features believed to be possessed by your friend (long blond hair) while eliminating irrelevant features (short or dark hair).

The relation of bottom-up processing to top-down control can be illustrated by considering how we come to recognize the identity of a visual object, such as an apple. By way of connections from feature-specific circuits in the primary visual cortex, shape and color information begins to build up in secondary areas, where simple representations are combined to form more and more complex ones in later stages in the temporal lobe, including the medial temporal lobe. Such integration of sensory representations with memory as the processing sequence unfolds allows recognition of the object as an apple.

But these bottom-up processes occur relatively slowly, and are supplemented by a faster process that involves executive functions of the prefrontal cortex. For example, secondary visual areas connect with the prefrontal cortex, which connects back to secondary areas. These loops allow top-down executive attention to temporarily maintain selected representations in an active state in the secondary visual cortex while the object is being constructed via bottom-up processing. In this way, bottom-up processing is facilitated by narrowing the focus of processing. But in addition, the prefrontal cortex also receives inputs from multimodal areas that store conceptual representations and can use them to anticipate what the object identity is likely to be, and thereby facilitate pattern completion of that identity in the secondary areas faster than would be possible by bottom-up processing alone. In some cases, such top-down processing may be necessary. Charan Ranganath offers the example that a glass of apple juice and a glass of beer (once it settles) can look very similar. If you are visiting your child in a nursery school, however, you do not see beer, while if you are in a bar, you do.* Recent studies suggest that even the primary visual cortex is influenced by top-down control.

Missing from this tidy account is the fact that the prefrontal cortex is not a unified, undifferentiated mass. It has a number of partitions that are also called upon to do cognitive work on a situation-by-situation basis (figure 49.4). Studies by Deepak Pandya, Michael Petrides, Helen Barbas, Joel Price, David Lewis, and Liz Romanski have shown intricate patterns of connectivity that link these various areas, enabling them to integrate sensory, memory, and internal milieu body signals into complex cognitive representations.†

Within the lateral prefrontal areas, a gradient of processing from the posterior to anterior end exists, with processing becoming less stimulus-specific and more abstract in progressively more anterior regions. The gradient is defined by the bottom-up inputs to the prefrontal cortex. The more posterior areas of prefrontal cortex primarily receive inputs from unimodal secondary sensory areas. Intermediate regions, like the dorsal lateral and ventral lateral prefrontal cortex, receive a combination of unimodal and multimodal inputs; these areas exert top-down control over their posterior inputs. The most anterior region, the frontal pole, only re-

---

* Paraphrased from a conversation with Charan Ranganath at a conference on June 9, 2018, at Duke University.

† I greatly appreciate helpful suggestions about prefrontal cortex anatomy from Liz Romanski, Helen Barbas, and Roozbeh Kiani.

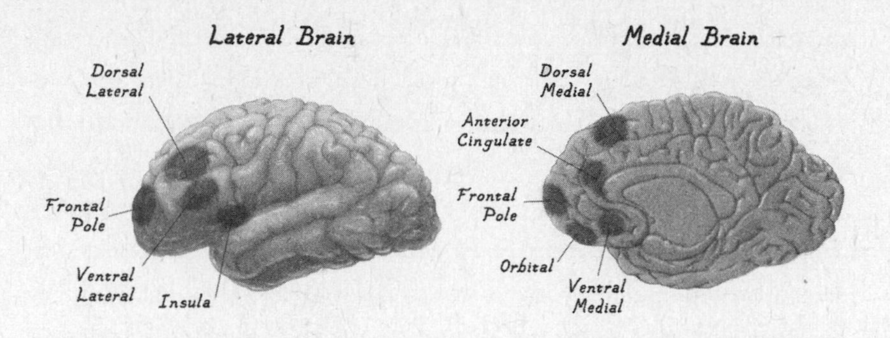


**Figure 49.4:** ***Key Areas of Prefrontal Cortex Implicated in Cognitive Processing***

ceives inputs from multimodal convergence zones, and creates the most abstract conceptual representations in the brain; it allows maintenance of long-term goals for future planning, and contributes to reasoning and problem solving. It interacts with the dorsal and ventral lateral prefrontal regions, and these together allow executive control over both unimodal (sensory) and multimodal (conceptual) processing in posterior areas, as well as control of deliberative behavior by way of connections to the motor cortex.

On the medial side, the anterior cingulate, orbital, ventromedial, and dorsal medial cortical areas receive inputs from medial temporal lobe memory circuits and also from subcortical areas that process body signals, such as the amygdala and hypothalamus. The insula region, buried deep between the lateral and medial cortex, also has similar inputs. These prefrontal regions are interconnected and also connect with lateral prefrontal areas. The medially located anterior cingulate cortex also contributes to executive functions, especially attention and monitoring of information processing itself; perhaps a more ancient region was a precursor to executive functions of the late-evolving lateral prefrontal areas.

The situational recruitment and coordination of prefrontal components likely involves executive functions of lateral prefrontal areas and anterior cingulate regions as they engage in top-down control of bottom-up processing, determining what will be focused on and how it will be processed. However, it is possible that some degree of auto-organization of prefrontal components occurs on the basis of what is being processed, with executive action by the frontal pole and dorsal lateral prefrontal cortex stepping in to adjust the balance on the basis of long-term and immediate goals, respectively.

Other than language processing, much of this description of working memory circuits applies in a general sense to all primates, not just humans. And some of it applies to other mammals as well. But the ability to use complex pattern processing and schema in top-down control over perceptual or memory systems, though present in all mammals, varies considerably. And, as hinted at earlier, the cognitive capacities of other mammals, including apes, fall short in comparison to humans. The next chapter explores how these differences between humans and other animals, especially involving the prefrontal cortex, play out in neural circuits.

*Chapter 50*

# REWIRED AND RUNNING HOT*

Although the cerebral cortex is part of our mammalian heritage, the human version of it has unique features that set it apart from those of other mammals, including apes and other primates, and help account for distinct human cognitive capabilities. And the area of the human cortex that differs most is the prefrontal region.

The prefrontal cortex first acquired fame as a result of the case of Phineas Gage. This nineteenth-century railroad worker suffered damage to his prefrontal cortex when an iron rod penetrated his skull during an explosion. As a result, Gage's ability to reason and make decisions was severely compromised. Decades later, the Russian psychologist Alexander Luria, studying soldiers with head injuries in World War II, confirmed that prefrontal cortex damage produced deficits in attention, thinking, and planning. With the cognitive revolution, these capacities came to be thought of as being associated with working memory. And indeed, studies of working memory in humans and monkeys have implicated the prefrontal cortex in the integration of information, the temporary storage of representations, and their control of thought and action.

Because human working memory capacities clearly outstrip those of other primates, some scientists have used evidence about the large size of the prefrontal cortex in humans to explain the cognitive differences (figure 50.1). When measuring brain size differences between animals, the most widely used calculation in recent times has not been absolute volume

* This title comes from Todd Preuss's article, "The Human Brain: Rewired and Running Hot."

or weight, since brain size increases as body size increases. Instead, most researchers use brain size relative to body size. This is viewed as an estimate of how much computing power is left after all the more routine body functions of the brain (sensing and movement, for example, which take up a significant amount of real estate in the brain) are accounted for. The typical finding that the prefrontal cortex in humans is relatively larger than that of other primates was thus consistent with the cognitive advantages that humans seem to have. More recent findings, though, based on more precise measures, suggest that the human prefrontal cortex is actually about as large as expected for a primate of our size. But there is more to the story.

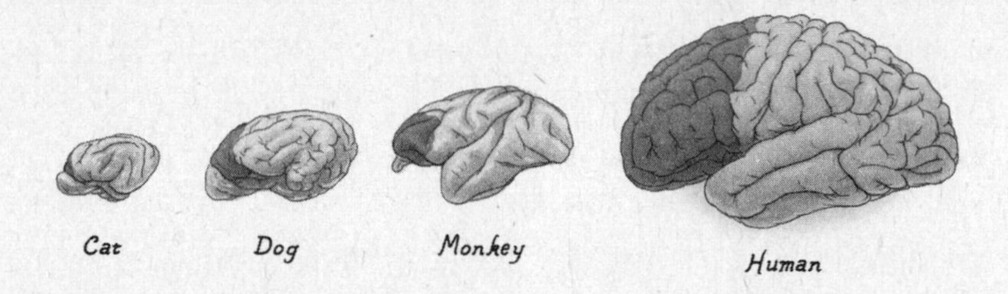


**Figure 50.1:** ***Size of Prefrontal Cortex Across Mammals***

Brain size measurements, whether they account for the entire brain or a single region, are easy to make, but are crude. As Todd Preuss, a leading thinker on this topic, points out, the cognitive functions of a particular brain area depend not so much on its overall size as on the internal organization and the cellular functions within it.

As noted in the previous chapter, the prefrontal cortex is not an undifferentiated region. One important difference of relevance here is between neocortical and paleocortical tissue (refer to figure 38.3). Neocortical tissue, you'll recall, consists of six layers. Paleocortical tissue has fewer than six layers. And a key layer missing in the paleocortical tissue is layer 4. This is significant because layer 4 of neocortical tissue in primates has a unique cell type—granule cells.

Key areas involved in working memory, and thus deliberative cognition, including the dorsal and ventral lateral prefrontal cortex and the frontal pole, are the prime examples of areas containing granule cells. The frontal pole is especially interesting. It is believed to be the area that most distinguishes the human brain from that of apes (figure 50.2), and Etienne Koechlin has described it as "a functional 'add-on' at the apex of the hierarchy of lateral prefrontal processes." He goes on to say it "enables cognitive branching; that is, the ability to put on hold an alternative course of action . . . requiring simultaneous engagement in multiple options that are not organized into a pre-established superordinate plan, such as reasoning, problem-solving and multitasking." Such cognitive branching is closely related to hierarchical relational reasoning. Goals set in the frontal pole can then influence the dorsal lateral region, which connects to the motor cortex in the control of deliberative behavior.

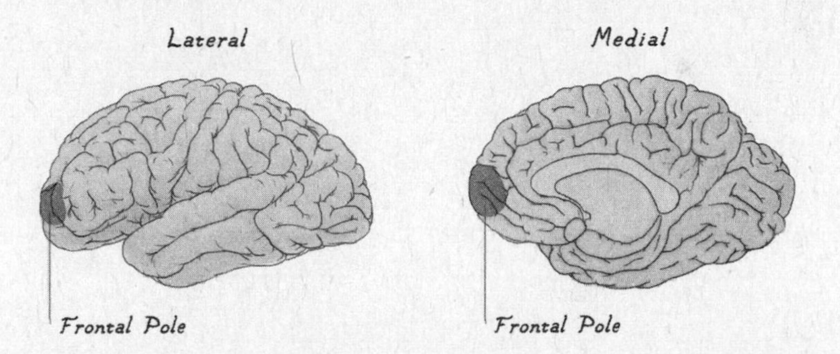


**Figure 50.2:** ***The Frontal Pole—A Unique Region of the Human Brain***

Most mammals have mainly paleocortical prefrontal areas, lacking layer 4 granule cells. Some large-brained mammals have lateral neocortical prefrontal areas that lack layer 4 granule cells. Only primates have granule cells in layer 4 of the prefrontal neocortex, which make possible unique patterns of between-layer processing. And recent studies by Earl Miller have shown that between-layer interactions in the granule prefrontal cortex are important in coordinating top-down and bottom-up processing.

While many features of the prefrontal cortex are shared by humans and other primates, the human prefrontal cortex has novel features not present even in apes—for example, unique spatial arrangements of cells and unique patterns of connectivity within and between cell layers. Moreover, neurons in the human prefrontal cortex are more strongly interconnected with neurons in other cortical areas than in apes and other primates (figure 50.3). For example, in humans and other primates, the various areas of the prefrontal cortex are highly interconnected with one another, as well as with other multimodal areas in the parietal and temporal lobes. However, in humans, greater connectivity exists between prefrontal and parietal and temporal areas, and also within the prefrontal cortex itself. Additional findings show novel patterns of gene expression in the human prefrontal cortex, especially in relation to energy metabolism and synapse formation. These various findings, and other observations described below, led Todd Preuss to describe the human prefrontal cortex as "rewired and running hot."

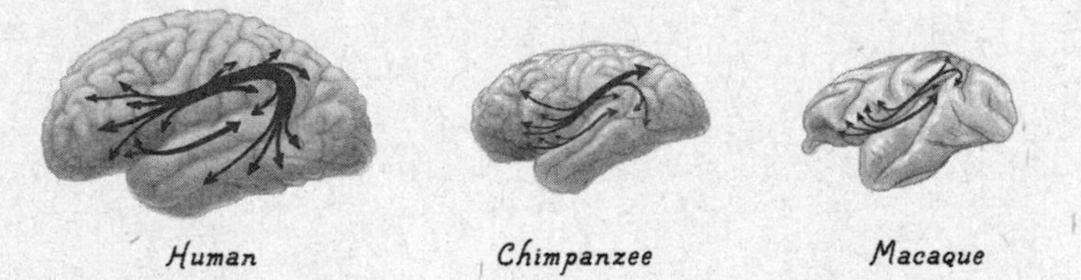


**Figure 50.3:** ***Greater Cortico-Cortical Connectivity in the Human Brain***

Some of the novel aspects of human cognition are prone to malfunction in conditions such as autism and schizophrenia. These cognitive traits and their neural underpinnings may not yet have been fully tested by natural selection for their contributions to fitness, and may be more susceptible to genetic perturbation than older, more established traits. Humans are, in general, recent members of the family of life and are a work in progress.

It was once thought that the two main areas in the human brain associated with language were unique to humans. While both Broca's and Wernicke's areas are now known to exist in apes, one of the most striking differences between human and ape brains is the pattern and strength of connectivity between these two areas in humans. Of note is the fact that both areas are interconnected with multimodal areas in the parietal, temporal, and frontal lobes, and contribute to our all-important ability to do mental work using verbal information. And as noted earlier, the ability to use language in hierarchical relational reasoning (cognitive branching) depends on working memory executive functions of the prefrontal cortex.

Language probably evolved in steps in early humans, but it surely also built upon basic cognitive processes that all mammals have, such as the ability to focus attention on external stimuli, to form memories about these stimuli, and to use these memories as internal representations to guide behavior in the absence of the actual stimulus. These various capacities may have been foundational to the emergence of working memory in mammals. In primates with bigger brains, and novel areas of the prefrontal cortex (relative to other mammals), working memory presumably got a significant power boost, allowing primates to hold more information in mind longer while deliberating about possible actions. With the arrival of humans, with even bigger, hotter brains, and especially bigger, hotter prefrontal cortices, cognitive capacities further increased.

The unique features of the human multimodal cortex would seem to go a long way toward establishing a foundation for understanding differences in cognitive capacities between humans and other primates, and primates versus other mammals. But let's not forget that culture also evolves, and that some of what makes us different from apes has occurred through cultural evolution, above and beyond the gene-based neurobiological evolution that enabled our ancestors to develop and sustain our brand of culture.

In celebrating our wonderful achievements, we have to keep in mind the long history that got us here. Yes, the human brain Bauplan differs from that of other animals, even closely related ones. But the brain Bauplan of every species differs in some way from that in the species from which it evolved. And, as is usually the case, overall similarities outweigh differences between the brain of new versus ancestral species. Human brains are no exception.

Differences, while important in defining a species, do not endow some with greater value than others in the vast scheme of life. We may prefer the kind of life we lead, but in the end there is no scale, other than survivability, that can measure whether ours is a better or worse kind of life, biologically speaking, than that of apes, monkeys, cats, rats, birds, snakes, frogs, fish, bugs, jellyfish, sponges, choanoflagellates, fungi, plants, archaea, or bacteria. If species longevity is the measure, we will never do better than ancient unicellular organisms.

As I have noted, scientists have to be vigilant in resisting both anthropocentrism and anthropomorphism. Navigating this terrain can be tricky, as we'll see in the next section on the topic of consciousness, a treacherous conceptual minefield.

# Part 12

# Subjectivity

*Chapter 51*

# BEING THERE*

I'm looking out a window at trees and other plants, and people walking around, some with dogs. The humans are attired in various styles of clothing, and the animals sport their fur. The grass is green, with the earth showing through in spots. There is a river nearby, and tall buildings across the way. The sky above it all is mostly blue, with patches of scattered, light-gray clouds.

My conscious mind, courtesy of my brain, seems to see these things, these meaningful objects, as "out there." Although the raw materials of what I perceive exist in the world, these are not what I experience when I "see" objects.

The perception of meaningful objects does not simply occur by one's brain building up sensory representations from the external raw materials. Also required is memory, which enables us to know what we are seeing when we see trees, dogs, rivers, buildings, clouds, and so on. This inner awareness of external stimuli—the knowledge of what is being perceived—is a prime example of what is called conscious experience.

Consciousness allows us not only to experience the present, but also to imagine the absent. We can, through consciousness, reminisce about our past and anticipate our future. Not just "the past" or "the future," but experiences in our own particular past and possible future.

Envisioning one's potential future is a particularly important skill, as it makes it possible to transcend the survival options bestowed upon our

* Thanks to Richard Brown for comments on chapters in this section.

species by natural selection, and/or instilled in an individual by goal-directed instrumental learning—both of which are based on behavorial strategies that were successful in the past. Future thinking about one's self is something humans do with ease, but, as we'll see in later chapters, has been difficult to demonstrate unambiguously in other organisms. It requires not only nonconscious deliberation using mental models and prospective cognition, which some animals can do, but also requires the ability to be reflectively aware of one's self as the agent of the forecast, and also as part of the subject matter. This may well be a human specialty.

Despite the central importance of consciousness to human mental life, psychology has had a complex relationship with it. The science of mind was officially born in the late nineteenth century. The first psychologists were introspectionists and prized consciousness, while behaviorists later banned it from the field. Cognitivists, upon dethroning behaviorism, kept consciousness at arm's length, and seldom touched it in the early days, mindful of the problems it had posed.

Meanwhile, though, consciousness had not been a taboo topic in the medical and biological sciences. Neurologists had long dealt with disturbances of consciousness in coma patients, and the discovery of the reticular activating system in the 1940s began to provide an account of how states of sleep and wakefulness are controlled by the brain. Research on other patients, such as those with epilepsy, also led to insights about consciousness as a mental state. By the 1950s, a field called neuropsychology was emerging, in which psychological consequences of brain injury were being studied using experimental methods. Its findings ultimately attracted the attention of mainstream cognitive scientists and neuroscientists, as well as philosophers, and paved the way for modern scientific research on consciousness.

For example, in the 1960s, Michael Gazzaniga and his PhD adviser Roger Sperry teamed up with a neurosurgeon and made remarkable discoveries about consciousness in so-called split-brain patients. These are patients with severe epilepsy in whom the axon bundle that connects the two cerebral hemispheres is surgically sectioned in an effort to control their seizures. People with typical brains can name common objects that appear anywhere within their visual field. Because language is usually controlled by the left hemisphere, split-brain patients are able to give verbal reports about information presented to the right half of visual space, which is preferentially seen by the left hemisphere, but cannot name stimuli in the left half of visual space, which is seen by the right hemisphere (figure 51.1). They can, however, respond nonverbally to the stimuli seen by the right hemisphere, by pointing toward or grabbing objects with the left hand, which is preferentially connected to the right hemisphere. Similarly, when blindfolded, these subjects can name objects placed in their right hand, but not objects placed in their left hand. Such findings showing that conscious experience can be isolated in a specific part of the brain by cutting axonal connections provided compelling evidence for something that scientists long believed but could not prove: namely, that consciousness depends on the brain's neural circuits, and is not, as Descartes had said, associated with a separate, nonphysical, soul.*

I was fortunate to be one of Gazzaniga's graduate students in the 1970s,

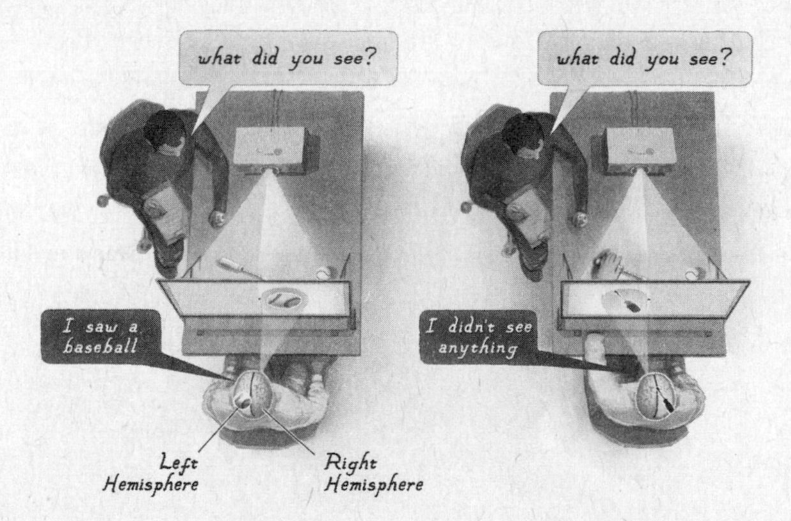


**Figure 51.1:** ***Classic Split-Brain Findings***

* A new study, "Split Brain: Divided Perception but Undivided Consciousness," shows that decades after surgery some of the more dramatic aspects of the disconnection dissipate, perhaps because of plasticity reflecting the brain's ability to compensate for the consequences of the surgical effects.

when a new group of patients became available to study. My PhD thesis focused on the question of what the left hemisphere makes of responses generated by the right hemisphere. After all, from the point of view of the left hemisphere, responses coming from the right were generated nonconsciously. But rather than being surprised by these responses, the left hemisphere took them in stride, and simply wove a tale that made the responses make sense. For example, we had a patient who could read text in both hemispheres but could speak only via his left. We flashed "Stand up" to the left visual field (in other words, to the right hemisphere), coaxing the person to stand up. We then asked him, "Why did you stand up?" He (or more accurately, his speech-equipped left hemisphere) said, "Oh, I needed to stretch." This was pure confabulation, since the left hemisphere was not privy to the information that instructed him to stand up. We observed many such instances of these confabulations.

To explain these findings we called upon the social psychologist Leon Festinger's theory of cognitive dissonance. The theory proposes that mismatches between what one expects and what happens create a state of inner discordance, or dissonance. Because dissonance is stressful, it demands reduction in order to meet the human need for cognitive equilibrium (figure 51.2). Our proposal was that when our split-brain patient became aware that his body produced a response that he did not initiate, dissonance resulted, and the confabulation of a reason why the response occurred was a means of reducing dissonance. Today, "postdecision ratio-

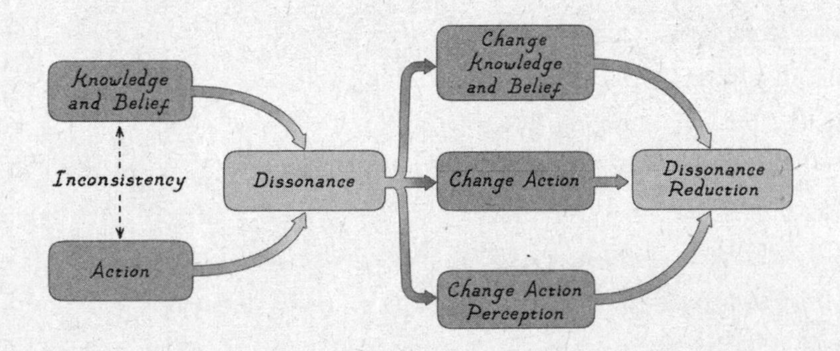


**Figure 51.2:** ***Cognitive Dissonance Theory***

nalization" is an active research topic that examines how people retroactively justify their decisions and actions in life.

A second group of patients that helped jump-start consciousness research suffered damage to the visual cortex of the right hemisphere, typically caused by a stroke. Patients with this kind of brain pathology are essentially blind to stimuli in the left half of their visual space. In pioneering research in the 1970s, Larry Weiskrantz showed that although the patients claimed not to see stimuli presented in this blind spot, they could accurately reach for objects located there, and could perform other behaviors that showed that the stimuli were, in fact, perceptually registered by the brain. Specifically, the stimulus was reaching systems in the right hemisphere that control behavior but that did not support conscious perception. This phenomenon was dubbed "blindsight"—the ability to respond behaviorally to visual stimuli that could not be reported on verbally.

A third group of patients had undergone surgical removal of the hippocampus and other areas of the temporal lobe. Like split-brain surgery, this was done to help relieve uncontrollable epilepsy. Although most of the patients were operated on in the 1950s, the full implications for consciousness of these treatments unfolded slowly over the next two decades. In initial studies performed by the psychologist Brenda Milner, the patients were found to have suffered a severe form of amnesia. At first, it seemed that a general memory deficit, a so-called global amnesia, was present. But later work by Milner and Suzanne Corkin determined that the patients retained the ability to learn and remember how to perform difficult motor tasks (for example, drawing objects while looking at their reversed reflection in a mirror). Over time, other examples of spared memory were identified, and it became clear that in addition to motor skills, the patients could also learn procedures and form habits, and could develop Pavlovian conditioned responses. Extrapolating across these findings, Larry Squire and Neal Cohen proposed in 1980 that the memory deficit resulting from temporal lobe damage was limited to conscious memory. For example, although the patients could learn motor skills, they could not consciously remember having recently acquired the skill. Conscious memory came to be referred to by the designations "explicit" or "declarative," and nonconscious memory as "implicit" or "procedural."

Across all three patient groups, the pattern of results was thus similar in revealing profound dissociations between information processing that controls behaviors nonconsciously and processing that underlies conscious, reportable experience. Thanks to these early studies of neurological patients, research on consciousness is thriving today in psychology and brain science.

*Chapter 52*

# WHAT IS IT LIKE TO BE CONSCIOUS?

Over the past several decades, consciousness has become an exciting research topic in cognitive science, and in its stepchild, cognitive neuroscience. In any research project, if you want to study something scientifically you have to design experiments that control for alternative possible explanations. In the case of consciousness, you have to separate information processing that occurs consciously from processing that takes place nonconsciously. And the standard way to do this, as in the patient studies described in the previous chapter, is to contrast situations in which people are aware of a stimulus with situations in which they are not.

A key technique used in such studies of healthy brains is subliminal stimulation. If you present a picture of a common object for a full second or more, study participants can usually tell you what they saw. As you decrease the amount of stimulus exposure time, the ability to give a report weakens, until at a 20–30 millisecond presentation, pretty much everyone denies having seen a stimulus, and certainly can't tell you what it was. But then the question is, How do you know that the presentation wasn't simply too fast to be registered by the brain at all?

The way to evaluate this is to obtain both verbal and nonverbal reports. For example, as in split-brain and blindsight research, you can have a subject select which of several pictures matches the subliminally presented stimulus. Another approach is to pair a stimulus with electric shock, and then record heart rate or some other autonomic nervous system response while presenting the stimulus subliminally. The usual finding is that when the shock-paired stimulus is later presented, autonomic nervous system responses result (if a stimulus without an association with the shock is presented, no such changes are seen). Although the subject is unable to give a verbal report about the subliminal stimulus, the nonverbal (autonomic) responses show that the brain processed the meaning of the stimulus. Subliminal stimulation thus eliminates consciousness of, but not the ability to perceive and respond to, visual stimuli.

Research on subliminal stimulation caught the attention of the advertising industry in the 1950s. Marketers had long recognized the value of invoking sexual, patriotic, and other highly charged forms of symbolism in persuading people to purchase products. Advertisers also knew that when consumers were aware of that influence, they could guard against it, and some turned to subliminal stimuli as a way to catch consumers off guard by directing campaigns to their unconscious minds. A theater in New Jersey, for example, reportedly used briefly flashed messages, such as "Drink Coke" or "Eat Popcorn," to increase traffic at the concession stand. Vance Packard's 1957 bestselling book *The Hidden Persuaders* brought subliminal and other forms of subversive advertising subterfuge into public awareness.*

Meanwhile, back in academia, research on subliminal stimulation fell out of favor when critics argued that the studies were not as effective as was assumed in demonstrating that stimuli truly bypassed consciousness—little bits and pieces of information might actually be slipping through in even these brief flashes. Later, however, new approaches arose that more effectively ruled out such leakage. For example, a common approach currently is to use "backward masking," a procedure in which a briefly presented stimulus is immediately followed by another stimulus (the mask) that further suppresses conscious access to the target stimulus. Neverthe-

* In recent years, subliminal methods have also been used in political campaigns. In the 2000 presidential race, an ad sponsored by the Republican campaign of George W. Bush showed a black screen while the voice-over criticized Democratic health policy, and especially that of candidate Al Gore. During the voice-over, the word *rats*, which is part of "Democrats," appeared in large letters for a few milliseconds. When this was discovered, the Bush campaign attributed it to a mistake. Also related is the recent story of Cambridge Analytica, which is accused of using data mining to unwittingly manipulate voter behavior in the election of Donald Trump. According to *The New York Times*, Cambridge Analytica was an offshoot of another firm, Strategic Communication Laboratories, which offered military and political groups in third-world countries "dirty tricks" (such as subliminal manipulation) as a way to achieve their goals.

less, critics continue to contend that some information may have leaked through, and ever more stringent criteria are demanded. As a result, the phenomena that do pass the tests for demonstrating nonconscious processing end up being very simple psychologically. Because nothing very sophisticated can therefore be studied, some claim that the nonconscious is "dumb." But, as noted by Anil Seth, these tests, by virtue of their restrictive design, "undersell the richness of real-world phenomenology."

The entire field of consciousness research is based on the assumption that consciousness is something that has to be generated. We are not conscious of everything that our brain does. So, by definition, all the information leading up to conscious experience is nonconscious. Indeed, as Sigmund Freud said, consciousness is only the tip of the mental iceberg. But even Freud did not appreciate the full extent of the cognitive unconscious.

Philosophers have long made the point that conscious experiences have a certain subjective quality to them. To emphasize this quality, the term *phenomenal consciousness* is often used. The reason a special term is needed is because the word *consciousness* has other meanings. For example, as we have seen, it is sometimes used to distinguish states in which one is awake, alert, and able to respond appropriately to stimuli, as opposed to being asleep, deeply anesthetized, or in a coma. When I use the term *consciousness*, I will be discussing the phenomenal kind, and assume it is a product of the brain.

But not all proponents of phenomenal consciousness agree with this kind of physicalist explanation.* Thomas Nagel, whose famous paper "What Is It Like to Be a Bat?" jump-started contemporary interest in phenomenal consciousness, is a dualist. He believes that the subjective qualities, sometimes referred to as "qualia," that define "what it is like" to be in a conscious state do not have a physical basis at all. Nagel thus rejects the idea

---

* Physicalists treat the mind as a part of the material world and assume it is subject to the laws of physics. Many physicalists assume that the brain is the physical basis of the mind. But some go in the direction of panpsychism, proposing that consciousness extends throughout the physical universe. And some physicalists who accept the importance of the brain reject notions like consciousness. These eliminitivists argue that science will replace this quaint conception with a more precise description.

that phenomenal consciousness can be understood scientifically. But as Anthony Jack and Tim Shallice argue, "It is precisely because we know 'what it is like' to be in certain mental states that we are able to bring this evidence to bear . . . on theories of consciousness." In particular, they contend, introspective evidence provides the basis for a scientific account of phenomenal consciousness.

The most common way to turn introspection into data is by obtaining a verbal report from a conscious subject. Such a report can reflect either of two forms of introspection. One is the result of active scrutinizing; the other is passive noticing. The virtues and limits of verbal reports will be discussed in detail later, in the context of animal consciousness; for now, we'll proceed with the assumption that it is an extremely valuable, if imperfect, tool for studying human consciousness.

The findings from functional MRI have been very clear: when stimuli are reportable, areas of the visual cortex and areas of the general cognitive cortical network that underlies working memory are activated, especially areas of the prefrontal cortex (and also the parietal cortex in some instances). But when a verbal report cannot be given, only the visual cortex is activated (figure 52.1). For many researchers, myself included, such results indicate that in order to have a phenomenally conscious and verbally reportable experience of visual stimuli, sensory processing in the visual cortex has to be further processed by cognitive control networks underlying working memory.

Consciousness research is a vibrant and contentious area with a variety of competing theories and approaches (table 52.1). My commitment to a

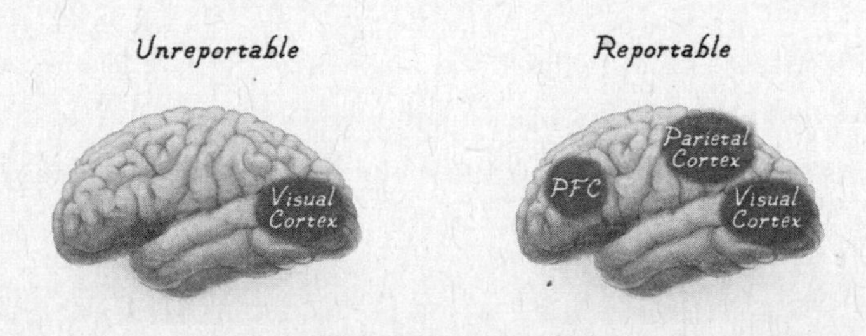


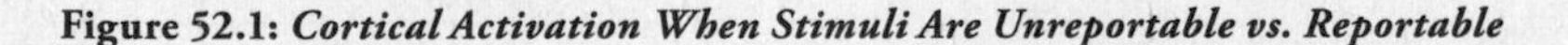
**Figure 52.1:** ***Cortical Activation When Stimuli Are Unreportable vs. Reportable***

cognitive approach has its roots in my experiences with the split-brain patient described earlier, and especially the confabulated verbal reports generated by his left hemisphere about behaviors executed by his right hemisphere. One possibility is that he was simply being deceptive—making up something he knew was not accurate. But he—specifically, his language-equipped left hemisphere—seemed to truly believe what he was reporting. It seemed, instead, that he was deceiving himself.

Table 52.1: The Foci of Some Contemporary Physicalist Theories of Consciousness

| Theories |
| --- |
| Attended intermediate representations (Jesse Prinz) |
| Attention schema (Michael Graziano) |
| Attentional amplification (Michael Posner) |
| Autonoetic consciousness (Endel Tulving) |
| Dissociable interacting systems (Daniel Schacter) |
| Dynamic core (Gerald Edelman, Giulio Tononi) |
| First-order representations (Ned Block, Victor Lamme) |
| Global workspace (Bernard Baars) |
| Global neuronal workspace (Stanislas Dehaene, Lionel Nacchache, Jean-Pierre Changeux) |
| Higher-order representations (David Rosenthal) |
| Higher-order representation of a representation—HOROR (Richard Brown) |
| Integrated information (Giulio Tononi, Christof Koch) |
| Microtubules (Roger Penrose, Stuart Hameroff) |
| Operating system (Philip Johnson-Laird) |
| Hierarchical predictive inferences (Karl Friston, Andy Clark, Anil Seth) |
| Social interactions (Chris Frith, Uta Frith, Nick Shea) |
| Supervisory executive system (Tim Shallice) |
| Verbal interpreter (Michael Gazzaniga) |
| Time-locked multiregional retroactivation (Antonio Damasio) |
| Working memory episodic buffer (Alan Baddeley) |

Plato said self-deception is the worst kind. But we had the idea that maybe this was not such a bad thing. The patient was, it seemed, simply calling upon a cognitive device that people use all the time in an effort to protect the unity of the self by reducing cognitive dissonance.

Our conscious mind is vain. It believes it is where the psychological action is. But we are more like a driver behind the wheel of a Tesla, where we can take control if needed, but the rest of the time we can consciously think about something else.*

If, indeed, a good deal of brain function occurs outside of awareness, then our conscious minds are somewhat in the dark about some of that action. For example, psychologists have demonstrated time and again that one thing that is often missing from conscious awareness is the reason or motivation for why a particular behavior was produced. We know what we did but not why. To maintain a sense of organismic unity in the face of such disunity, consciousness must therefore have some sophisticated way of rescripting one's history to account for responses that it did not intentionally will.

Most of the time, in the case of trivial actions like waving your hands while talking or shifting your posture in your chair while seated, you simply pay no attention to your body's movements. But sometimes nonconscious responses compel or require an accounting—for example, when your action is at odds with what you think about yourself, you can generate an explanation that rationalizes you to yourself; or if someone asks you why you responded the way you did, and you are embarrassed to say you don't know, you can simply assign ownership of the behavior to yourself. These narrative moves reduce dissonance and help maintain a sense of control and personal unity. As Daniel Dennett has noted, such narratives are a defense tactic, a way of defining and protecting our understanding of our self.

Coming at this from a completely different set of experiences, assumptions, and references, Jeff Tweedy, the front man of the indie-rock band

---

* Thanks to Milo LeDoux for suggesting this analogy.

Wilco, and one of my favorite songwriters, reached a similar conclusion in his memoir, *Let's Go (So We Can Get Back)*. He noted that our brains are wired to make sense of things by eliminating ambiguity. Tweedy makes sense of his own life by writing songs, which, of course, are narratives.

When Mike Gazzaniga and I were writing *The Integrated Mind* in the late 1970s, working memory was just emerging as a psychological construct. Had we known then what is known now about this topic, we would have been able to present a more detailed and nuanced cognitive account of what goes on when people generate narratives to account for their actions. Specifically, working memory is what makes possible the internal monitoring of events in the external environment, including one's own behaviors, and the garnering of internal information from memory, including schema and models, about one's self and the world, to generate an interpretation, a narrative, about those events and actions. Indeed, prefrontal working memory circuits have been implicated in producing narratives, schema, mental models, representations of the self, confabulations, and consciousness itself.

Because the ideas in this section are so dependent on those 1970s split-brain studies, I decided to do something I hadn't done in a long time. I opened up *The Integrated Mind* to see exactly what Gazzaniga and I had said in print. Here's one informative passage:

> Why did George suddenly find himself in bed with Molly [not his wife] in the first place? What is the mechanism for eliciting a dissonant behavior from the beginning? The behavior was clearly contrary to his existing (verbally stored) belief about such matters, and normally the verbal system can exert self-control. The reason we propose is that yet another information system with a different reference and a different set of values existed in George, but because it was encoded in a particular way, its existence was not known to George's verbal system and therefore was outside of its control. This other system wasn't known to the dominant verbal system until the day it grabbed hold and elicited a behavioral act that caused great consternation to his verbal system.

Once elicited, however, George's verbal system had no choice but to account for it and to adjust his verbal perceptions and guidelines for behavior in such a way as to take this newly discovered aspect of his personality into account. In this view, it is the verbal system that is the final arbiter of our multiple mental systems, many of which we come to know only by actually behaving.

This exercise confirmed my belief that my ideas about consciousness were crystallized in those first few moments watching the patient confabulate explanations, and then discussing the day's observations at the bar that night with Mike.

*Chapter 53*

# I WANT TO TAKE YOU HIGHER*

The nervous system is fundamentally a device that gathers sensory information about the world for the purpose of guiding behavior in the quest to survive. As a result, awareness of the external world is perhaps the most basic level of conscious experience. In an influential 1990 paper, Francis Crick and Christof Koch argued that, because so much was already known about sensory processing, and especially vision, progress in understanding the neural basis of consciousness might be best achievable by studies of visual perception.

As noted above, evidence from fMRI imaging studies in the 1990s showed that areas of the visual cortex and the prefrontal cortex are both activated when participants are able to verbally report on visual stimuli, but only the visual cortex is absent when stimuli cannot be reported. Today, the exact role of these different regions is the subject of much debate.

On one side, my NYU colleague Ned Block argues that perceptual processing in the sensory cortex is all that is required for phenomenal consciousness, and that what circuits in the prefrontal cortex make possible is cognitive access—the ability to introspect and give verbal reports, independent of the phenomenal experience. Because the visual cortex state is a representation of a physical event in the world, it is called a first-order state, and Block's theory is accordingly a first-order theory of consciousness (figure 53.1). Block made a case for his idea by citing a variety of findings showing that people process more information than they can

---

* Thanks to Richard Brown for comments on higher-order theory.

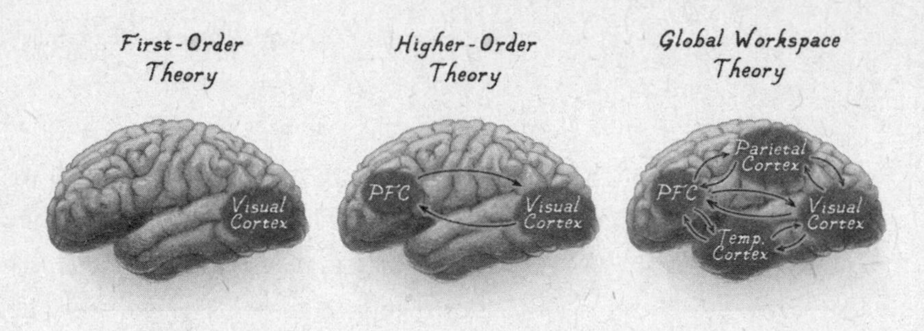


**Figure 53.1: *Three Theories of Conscious Visual Experience***

report on, and, in general, can respond behaviorally to stimuli that they cannot describe verbally.

Some, myself included, find this an odd notion, since it separates phenomenal consciousness from the ability both to know what you are seeing and to be able to report on that experience. For example, Nathan Giles, Hakwan Lau, and Brian Odegaard argued that findings such as those used by Block are really about nonconscious perceptual processes that control behavior rather than about an actual state of phenomenal consciousness. If you don't know something is there, you can't actually be having a conscious experience of it.

Cognitive theories, by contrast, assume that conscious awareness of a visual stimulus requires processing beyond the sensory cortex, and particularly by the prefrontal cortex. While there are numerous such theories, I will focus on two leading contenders here.

In David Rosenthal's higher-order theory (HOT), conscious awareness results when nonconscious first-order sensory information is cognitively re-represented, resulting in a higher-order state, which is a mental state about a lower-order mental state. The way to think about the difference between first-order and higher-order theory is in terms of the kinds of states involved: first-order theory focuses on a mental state that represents the world, while HOT adds an additional (higher) mental state that re-represents the sensory state. By definition, then, in HOT, the first-order state is a nonconscious one, and only becomes a conscious mental state with the help of a higher-order one. At the risk of oversimplification, the higher-order state makes conscious the lower-order state.

One implication of Rosenthal's HOT is that we are not conscious of the higher-order state itself, only of the lower-order state. To become conscious of the content of the higher-order state requires that it be re-represented by an additional higher-order state. For example, HOT assumes that when we consciously see a red apple it is because the prefrontal cortex makes conscious the visual cortex's representation of the visual properties of the object. But to be aware that you are having the experience, an additional higher-order state, possibly also involving the prefrontal cortex, is needed.

Another contemporary cognitive theory of consciousness is global workspace theory (GWT). First proposed by Bernard Baars, GWT assumes that specialized processing modules in posterior areas of the cortex (perceptual, memory, language) operate unconsciously, and vie to post information in the cognitive workspace of working memory. The information successfully posted there is then broadcast widely throughout the brain, making it available for use in thought and behavioral control. Consciousness, in this view, is equivalent to the global availability of disseminated information.

Stanislas Dehaene, Lionel Naccache, and Jean-Pierre Changeux reframed GWT in neural terms as a global neuronal workspace. In their model, executive attention selects what is posted in prefrontal working memory circuits, and also amplifies the processing of the selected information in the specialized modules, creating loops of processing between the specialized module and the workspace used in cognitive control of mind and behavior.

GWT and HOT both argue that low-level processors operate nonconsciously, and that some additional processing is required for consciousness. They also agree that prefrontal cognitive networks underlying working memory play an essential role in that process. But they diverge in describing how consciousness arises from this neural architecture. In GWT, consciousness of a perceptual state is equated with the global availability of the information. However, Rosenthal has pointed out that global availability does not differentiate what makes some states conscious and others not, since global broadcasting occurs in both kinds of states. Indeed, Baars himself noted that some additional kind of representation is necessary to create awareness of the content of the broadcast information, which sounds a lot like HOT.

GWT is appealing in many ways, but to me, HOT seems to have the edge since it explicitly attempts to account for how conscious content is actually experienced. But it, too, has its critics.

For example, given the assumption that a conscious experience depends on first-order state, how can HOT account for situations in which one's conscious experience *mis*represents the world—for example, if the object in the world is green but is experienced as blue (as in some cases of color blindness), or if one has the experience of a specific object, but no such object is actually present (as in hallucinations).

HOT enthusiasts Hakwan Lau and Richard Brown responded to the above criticism by citing findings from neurological patients with rare Charles Bonnet syndrome, a condition in which first-order sensory states cannot exist because of damage to the source of the first-order state, the visual cortex. The patients nevertheless were able to give reports of having vivid visual experiences (which were in effect hallucinations) of faces, objects, and geometric shapes. Lau and Brown suggested that first-order sensory states are therefore not absolutely necessary for conscious perceptual experiences. In other words, the higher-order state alone is sufficient. Brown calls this version of HOT "higher-order representation of a representation theory" (HOROR theory).

As we go forward, I will emphasize HOT, not because it has been shown to explain consciousness, but instead because it resonates with my long-standing views of consciousness, and I can envision ways that it might be leveraged to account for how complex conscious experiences involving emotions and memories come about in the brain. Next, I thus explore a neural account of HOT in more detail.

*Chapter 54*

# HIGHER AWARENESS IN THE BRAIN

Hakwan Lau and David Rosenthal offered the first formal neural account of higher-order consciousness in 2011. They built on the body of work that had shown that neural activity in the prefrontal cortex is correlated with the ability of study participants to give introspective verbal reports and make cognitive judgments about their perceptual experiences. To these correlational findings they added evidence that prefrontal damage, or functional inactivation of the prefrontal cortex,* disrupts conscious perceptual experiences. Most important, though, they offered specific suggestions about what the prefrontal cortex does in the context of HOT.

A key part of the model was its specification of areas of the prefrontal cortex that might constitute the core circuitry underlying higher-order perceptual awareness (figure 54.1). One of the regions they identified as potentially important was the dorsal lateral prefrontal cortex. This area has the requisite inputs from secondary visual areas that are needed to form higher-order representations of lower-order visual states. And given the well-established role of the dorsal lateral prefrontal cortex in top-down executive functions, it also has the control and monitoring capacities required to act back on and affect sensory processing. One limitation of their model is that while the dorsal lateral area receives some, it does not receive a full complement of sensory inputs. But this limit is easily overcome if we include the adjacent ventral lateral prefrontal area, which has

* Transcranial magnetic stimulation is a powerful, relatively new approach to temporarily and safely alter the function of specific brain regions in healthy humans.

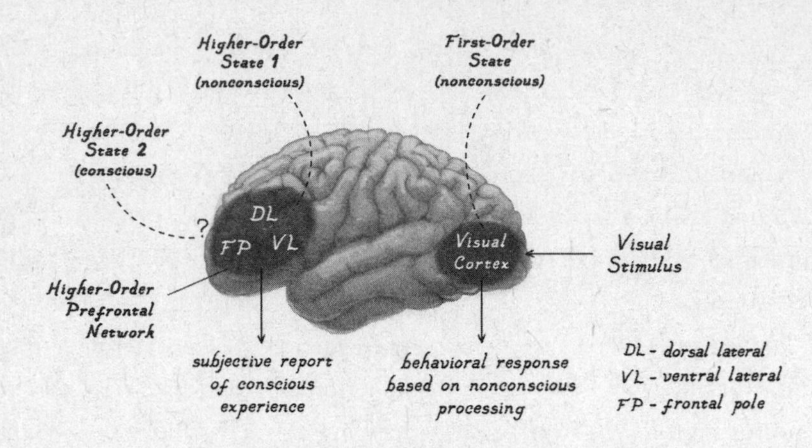


**Figure 54.1:** ***Lau and Rosenthal's Higher-Order Network of Consciousness in the Prefrontal Cortex***

connections from additional sensory modalities, and is also interconnected with the dorsal lateral area.

Also important is the fact that Lau and Rosenthal included the frontal pole as part of the prefrontal higher-order consciousness network. This region, as we've seen, has features that are not present in other animals, including other primates. It receives minimal if any unimodal sensory inputs, but it is massively connected with multimodal neocortical convergence zones, including the temporal pole and other neocortical areas that integrate sensory inputs and store semantic memories and concepts. The frontal pole is also interconnected with other prefrontal areas. With these inputs, the frontal pole is well suited for high-level conceptual processing and is, in fact, generally considered to have the greatest capacity of any brain area for conceptual processing.

Thus, dorsal lateral and ventral lateral prefrontal areas are higher-order anatomically with respect to posterior unimodal and multimodal processing areas, and the frontal pole is higher order with respect to all of these. Consistent with this processing hierarchy, as noted previously, task-specific subgoals are believed to be managed by the dorsal lateral area, while the frontal pole is thought to be more involved in representing long-term goals, and in cognitive multitasking and hierarchical reasoning.

Findings by Steve Fleming have shown that the frontal pole uses information from the dorsal lateral prefrontal cortex in making introspective judgments related to conscious experiences of sensory stimuli. Fleming concluded that the frontal pole contributes to a subject's ability to change her mind about what she consciously thinks and believes, and how confident she is in those beliefs. This capacity to construct a "mental big picture" would seem to be a key feature, and advantage, of human consciousness.

Sitting at the top of the prefrontal hierarchy, the frontal pole is thus well situated to play a key role in higher-order experience. It is able to monitor high-level information processing in the brain and to use its extensive capacity for conceptualization to initiate top-down executive control over processing in other areas, including other prefrontal areas.

Critics argue that prefrontal areas lack the kinds of fine-grain representations that could support the rich quality of perceptual experiences, but recent findings by Brian Odegaard, Lau, and other colleagues suggest otherwise. Using new, sophisticated methods of measuring brain activity, they obtained evidence that detailed content can actually be read out (decoded) in prefrontal neural activity signals.* Thus, the higher-order prefrontal network could conceivably contribute to the creation of rich phenomenally conscious perceptual content.

Critics of higher-order theory also argue that people with frontal lobe damage do not always lose the ability to have reportable conscious experiences. However, Odegaard, Lau, and Robert Knight have pointed out shortcomings of the criticisms. For one thing, the higher-order network is seldom completely damaged bilaterally. But very important is their observation that the critics actually misreported the boundaries of prefrontal damage, which in effect negates the critique. Finally, what is meant by "consciousness" and how "consciousness" has been measured has varied in the reports. That some kind of awareness is possible or missing when undertaking some tasks does not mean that all possible functions are present or missing.

---

* This 2018 study, "Can Perceptual Content Be Read-out from Prefrontal Cortex?," used a powerful new approach involving hyperalignment of fMRI signals within and between brains.

I will be arguing that the higher-order network uses schema in assembling higher-order conscious experiences. Specifically, schema, in my view, are nonconscious memory and/or conceptual representations that, when pattern-completed in response to momentary external or internal events, influence (if not dictate) the content of one's present conscious experience. This idea would help to explain how conscious experiences could come about in the absence of actual sensory input, as in Brown's HOROR theory (activation of the schema generates the content). But it is also compatible with models based on sensory input, such as Lau and Rosenthal's idea that top-down processes contribute to the experience of first-order states, and Lau's joint determination view of HOT, which proposes that the higher-order prefrontal representations index the relevant first-order content in sensory areas to allow that content to enter consciousness such that activity in higher- and lower-order networks "jointly determine" the experience.

In providing a clear and detailed neural account of how higher-order re-representations of sensory information might come about in the brain, Lau and Rosenthal did a great service to the field.

I have suggested some additions that, I think, enhance its explanatory power. But, even with these enhancements, the model, as it stands, is better at accounting for perceptual experiences in controlled laboratory settings with simple stimuli than for the kinds of complex conscious perceptual experiences you have as you move through the episodes of your life, and is even less adequate for accounting for how you experience your self and your emotions.

I believe that Rosenthal provided a clue about what else is needed when he argued that, contrary to popular opinion, his higher-order theory is not an account of how cognitive states make lower-order sensory states conscious. Instead, he says, the relevant lower-order states represent some combination of sensory and conceptual information.

Conceptual representations of objects and events in the world are, in effect, memories (including schema) about the ways things are similar to and different from other things. We acquire such concepts by the accumulation of information across multiple experiences in life. Given that sensory cortex areas are not pure sensory processors but also store information about objects encountered in the past and use that information to process present objects, they could provide the kind of sensory-conceptual merger Rosenthal had in mind.

While I am very keen on the idea that conceptual information plays a key role in consciousness, there are two reasons to think that sensory-conceptual representations involving secondary sensory cortex might not constitute the relevant conceptual representations, or at least might not be the only important ones. First, recall that rare Charles Bonnet syndrome patients have meaningful conscious visual experiences, despite having brain damage of secondary visual areas. Second, there is new appreciation for the idea that prefrontal circuits use conceptual information (including schema) stored as memories in multimodal neocortical areas and in the medial temporal lobe to do heavy lifting in determining, in top-down fashion, what we see when we have conscious perceptual experiences. As we explore next, conceptual expectations play an important role in determining our perceptual experiences.

## Part 13

# Consciousness Through the Looking Glass of Memory

*Chapter 55*

# THE INVENTION OF EXPERIENCE

Ewald Hering, a late-nineteenth-century pioneer in the study of the physiology of vision, wrote: "Memory connects innumerable single phenomena into a whole, and just as the body would be scattered like dust in countless atoms if the attraction of matter did not hold it together so consciousness—without the connecting power of memory—would fall apart in as many fragments as it contains moments." Around the same time, William James noted: "Whilst part of what we perceive comes through our senses from the object before us, another part (and it may be the larger part) always comes out of our own head." A century later, the Nobel Prize–winning immunologist turned neuroscientist, Gerald Edelman, captured the sentiment of both Hering and James when he described consciousness as the "remembered present." The memory neuroscientist Richard F. Thompson similarly proposed, "Without memory there can be no mind." In short, these authors are saying that what we see, think, and feel in a given situation depends on what we have experienced in the past—that we experience the present through the lens of memory.

Hermann von Helmholtz, another nineteenth-century pioneer in perceptual research, emphasized that "experience, training, and habit" influence our perceptions because they allow us to have "unconscious conclusions" about what is present that do not necessarily fully match with the sense data available. He used many examples, including optical illusions and phantom limb symptoms, to support what has since come to be called "unconscious inferences," or expectations that shape perceptions.

The idea that what we experience is shaped by memory-grounded expectations is sometimes referred to as perceptual set, a topic that came into the limelight of contemporary psychology in the 1940s and '50s through the writings and research of Jerome Bruner. One of the examples he wrote about is shown in figure 55.1. The item in the center is seen as the letter B reading down (which puts the central item in the context of letters) but as 13 when reading across (which puts it in the context of numbers). And in now-famous studies, he showed that when anomalous playing cards (for example, a red ace of spades or a black queen of hearts) were shown briefly to people, they failed to notice anything amiss. However, if told to expect the unexpected, they detected the anomalies, even with the brief exposures. He also brought sociocultural factors into the study of perception, showing that children from lower-income families, when asked to estimate the size of coins, viewed them as larger than did children from wealthier backgrounds.

Another classic example of how expectations affect experience is shown

**Figure 55.1:** ***Top-Down Influence on Perception***

on the right side of figure 55.1. Most people have trouble seeing a meaningful object in this image. But once told that it is a Dalmatian, many then subsequently recognize the dog—the stored conceptual knowledge of what to expect allows it to be seen. The psychologist Richard Gregory used such examples to argue that because neural processing compresses sense data so much, the brain has to reconstruct what is there using what Helmholtz called the "likelihood principle." That is, we use prior knowledge to unconsciously infer what is there.

In recent years, this general idea has received renewed attention in the form of what has come to be known as the predictive coding hypothesis. The essence of the idea is simple—that top-down unconscious predictions based on retrieved knowledge, or memories (which are called "priors" in this context), shape what we consciously see.

While there have been many versions of this hypothesis, one that has attracted considerable attention lately is Karl Friston's hierarchical active inference model. Friston proposes a complex mathematical account of the processes by which the brain makes predictions and updates expectancies, but we will focus on the big picture here. Figure 55.2 illustrates the basic idea in a highly simplified form that only includes interactions between the sensory cortex and prefrontal cortex. Bottom-up inputs reaching the prefrontal cortex trigger predictions about the state of the visual cortex. Differences between the prediction and the actual visual cortex state at the moment generate errors that effect changes in the prefrontal cortex, resulting in a new set of priors that improve prediction accuracy in the next iteration of the bottom-up/top-down dance. Iterations of this hierarchical process underlie the momentary inferred perception.

But also important to note is that similar interactions (not shown in the figure) are occurring between the sensory cortex and memory/conceptual processors, and between them and the prefrontal cortex (and also within each of the areas and their subareas). As a result, the prefrontal

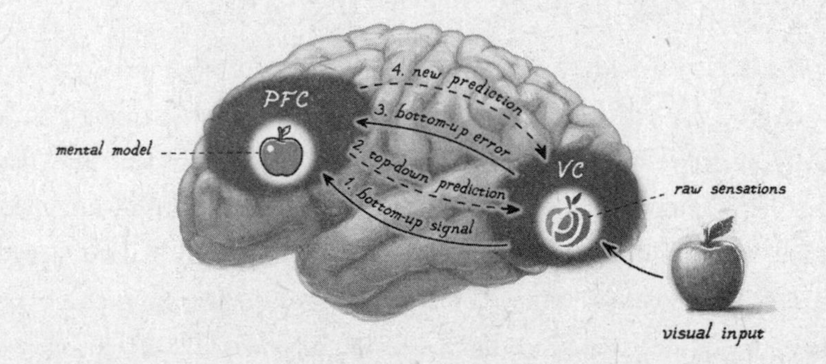


**Figure 55.2:** ***Hierarchical Predictive Coding: Perception Depends on Top-Down Predictions Updated by Bottom-Up Prediction Errors (intermediate steps not shown)***

cortex is not simply inferring what is present in the outside world based on sensory processing; it is also using memory/conceptual information, which, as noted in chapter 49, can pattern complete perceptual experiences faster than bottom-up sensory processing can.

Andy Clark and Anil Seth, both enthusiasts of predictive coding, characterized conscious perception as a "controlled hallucination," and Chris Frith described it as "a fantasy that coincides with reality." Lucia Melloni went further, writing, "The images that reach consciousness often bear little resemblance to reality." She quotes the Austrian scientist and philosopher, Heinz von Foerster, to support this point: "The world, as we perceive it, is our own invention."

Seth nicely summarizes the predictive coding approach: "Our perceptual experience—whether of the world, of ourselves, or of an artwork—depends on the active 'top-down' interpretation of sensory input. Perception becomes a generative act, in which perceptual, cognitive, affective, and sociocultural expectations conspire to shape the brain's 'best guess' of the causes of sensory signals." He goes on to quote Georges Braque, the French Cubist painter, who said, "Objects don't exist . . . except insofar as a rapport exists between them, and between them and myself."

One criticism of the predictive coding hypothesis is that it is just a fancy label for the traditional notion of perceptual set, with the supporting math being a kind of smoke screen that makes it hard to evaluate. Another is that it overvalues top-down processing and shortchanges bottom-up effects.

I tend to agree with the top-down enthusiasts, and view top-down effects on perception in terms of schema or mental models that guide pattern completion and separation. They can be said to provide the "priors" underlying the unconscious predictions/inferences that, in turn, complete the patterns from limited sensory cues. An obvious advantage is that unnecessary bottom-up computational work can be bypassed, saving brain energy and other resources. The downside, of course, is that predictions can be wrong. But if the system is constantly updating predictions based on error corrections, the errors are small and rapidly compensated for.

A potential problem with my schema hypothesis is that mental models generated by expectations are approximations. But this is easily dispelled. For example, while a spatial map in the brain lacks the minute details present in the Euclidian space of the "real" world, or even in a detailed map of it, it is good enough to guide navigation. In the same way, a working memory model of an object or scene is sufficient to enable us to see the object or scene when the present situation conforms to what we expect to see, given past experience. And when it does not conform, error correction kicks in, initiating new predictions, and updating of perception.

Clearly, when you "see" a painting or "hear" a song, your experience lacks information present in the source. That our conscious perceptions are missing details that are present in the sensory world reflects two limitations of sensory processing. The first is that our sensory systems are unable to capture all of the details present in the external environment at any given moment—there is just too much information to absorb. And second, regardless of the fidelity of sensory representations, the information that reaches awareness is further restricted by working memory's limited capacity. This is the price paid for having a cognitive system that can integrate information from many detailed processors to create complex abstract representations without having to have the neurocomputational power to recapitulate all the details in a given representational source. There can be no denying that our experiences are psychologically rich in apparent if not actual detail.

Not surprisingly, the prefrontal cortex is believed to play a key role in top-down predictive coding using mental models. It, more than any other brain region, has the capacity to generate and update complex conceptual representations of what we experience in a given scene at any one moment.

The idea that conscious states entail conceptual predictions based on priors (memories) related to lower-order sensory processes fits well with the theme I have been developing—that conscious experiences are higher-order states that depend on memory. This theme will follow us much of the rest of the book.

*Chapter 56*

# AH, MEMORY

Despite my repeated reference to the importance of memory in consciousness, it is important to note that memory has actually played a limited role, at best, in research on consciousness, which, as we've seen, has emphasized the role of sensory processing in perceptual experience. Consciousness, on the other hand, has, for some time, been front and center in the study of memory.

Memories that can be brought into conscious awareness and verbally described are, as we've seen, referred to as explicit or declarative memories (figure 56.1). These are distinguished from implicit or procedural memories, which do not depend on conscious awareness. The latter underlie behaviors learned as conditioned responses, habits, skills, or procedures.*

An important clarification needs to be introduced here. Although explicit memories are described as conscious, they are not literally conscious memories. That is, they are not consciously experienced until they are retrieved from storage and brought into working memory.

The psychologist Endel Tulving proposed an influential distinction between two kinds of explicit memories—semantic and episodic—which were discussed briefly in part 11. Here we will go into more detail about them, since they are key to my approach to consciousness.

Semantic memories concern facts—things you know about the world.

---

* Although recently there has been some debate about whether dependence on consciousness is the best way to divide memory psychologically for the purpose of defining memory systems in the brain (other approaches that have been proposed emphasize processes or components that cut across the implicit-explicit categories), my concern here is not with the question of how to categorize memory systems. It is instead with how conscious memories themselves come about.

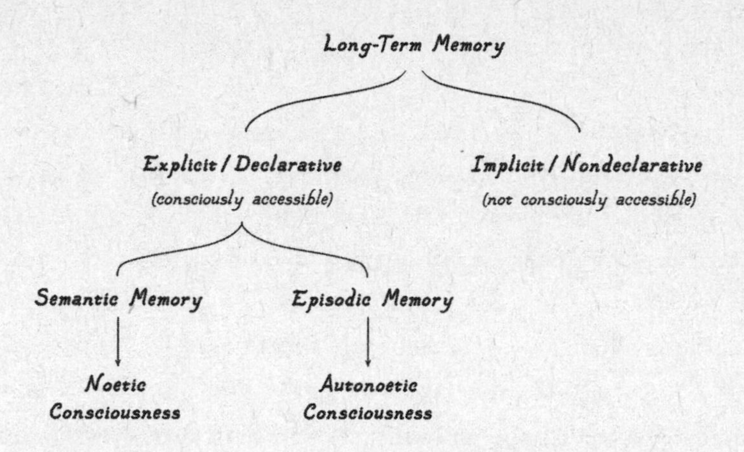


**Figure 56.1:** ***Conscious Experience of Explicit (Semantic and Episodic) Memory***

They are often acquired as the result of repeated experiences that form abstract representations of objects. Although not all apples are identical, over the course of many experiences with them you learn in a general sense what apples are—they are a roundish, often reddish, fruit about the size of a fist. But you also learn what they are not—they are similar to cherries in that they are both round, reddish fruits, but they differ in size and taste; both also differ from round red balls and marbles. Such semantic knowledge enables you to act appropriately when you encounter one of them—for example, to eat apples and cherries but not red balls. Semantic memories of object names allow us to label objects verbally, but semantic memories are first and foremost about providing object information independent of verbal labels—animals obviously have the ability to recognize objects and act appropriately toward them, despite lacking language.

Semantic memories are acquired through specific experiences, but they are not tied to those experiences in any defining way. Episodic memories, by contrast, are linked to the specific episodes from which they arise. Because an episode consists of a set of facts about a given event, episodic memories are often said to be defined by composite "what-where-when" representations. In addition, episodic memories include the experiencer in the experience, and are thus personal. You can form semantic memories about Italy by reading or hearing about it. But to have an episodic what-where-when memory of the country, *you* actually have to have been there.

Episodic memories are, by their very nature, autobiographical. But not all autobiographical memories are episodic. We accumulate factual (semantic) information about who we are as we go through life. For example, your knowledge of your name, and the name of your hometown, the school you attended, your first pet, and your favorite song or movie when you were a teenager are all semantic autobiographical facts that were acquired during specific episodes and used in later ones. So are your memories of your personal attributes, such as your abilities and skills, preferences, and psychological and behavioral tendencies. But when you remember a specific experience (what happened to you and when and where it happened) in which your skills, preferences, or tendencies were expressed, it is an autobiographical episodic memory that includes autobiographical semantic facts.

Tulving proposed a second very important distinction—that our conscious experiences of semantic and episodic memories depend on different kinds of conscious states (see figure 56.1). To be conscious of a semantic memory, he said, requires noesis, an awareness about facts based on stored internal representations of objects and events. By contrast, consciousness of episodic memory depends on autonoesis, an awareness of you, the experiencing person, as part of the experience. In noetic consciousness of semantic memory you are not personally a key part of the memory. The memories that underlie states of autonoetic consciousness, in contrast, are personal. They come with a "feeling of ownership"—the sense that they are memories about your own life.*

A key feature of autonoetic consciousness, according to Tulving, is "mental time travel." This process enables you to recollect the perceptual details, thoughts, and feelings that occurred during an event that you experienced at a specific time in your life (figure 56.2). Mental time travel also makes it possible to imagine your future—not just "the future," but your personal future.

* Tulving also described a third kind of conscious state, a-noesis, which was somewhat difficult to understand. In a conversation about this, he told me that what he meant by a-noesis is what most people might refer to as an unconscious state. I thus do not include a-noesis as a kind of conscious experience here.

Your memory that a restaurant called Trattoria San Lorenzo is located in a certain district in Rome involves noetic consciousness, while autonoetic consciousness underlies your ability to recall your experience of an enjoyable meal you had there, and to imagine yourself dining there again in the future.

**Figure 56.2:** ***Mental Time Travel***

To understand episodic memory as a conscious experience, we need to know something about the self. The acclaimed developmental psychologist Michael Lewis distinguished two aspects of the self—the machinery of self, and the self as a mental state. The machinery includes all of the biological requirements of being a self-sustaining organism, including a set of mutually compatible genes, immunological self-recognition, and a homeostatic mechanism that maintains self-regulating body functions. It also includes a variety of behavioral tendencies, often referred to as personality or temperament, that depend on either genetic influences or learning and that are expressed automatically—you don't have to consciously remember your core personality.

The machinery of self is established at an early age, and enables even an infant to respond in a "self-protecting" or "self-serving" way. Such responses are sometimes viewed as evidence of self-awareness, but I don't think they are, as they do not depend on the mental state of self. The mental state of self, according to Lewis, arises later, typically between eighteen and twenty-four months of age, as the child's brain attains cognitive wherewithal, including linguistic competence with personal pronouns—me, myself, I, and mine. The maturation of such cognitive tools equips the child to conceptualize the difference between self and other, and makes possible reflective consciousness, or self-awareness—in other words, autonoesis. In his book *Self Comes to Mind,* Antonio Damasio discusses a related view of the mental state self—the idea of the self-as-subject, as the knower, the "I," the "me." This is more than simply a mind capable of knowing when sensations (including body sensations) or images are present and exist. It is also capable of knowing that "I" exist, and the sensations and images exist within "me." Damasio refers to these as the difference between *self as subject* and *self as object.* Uriah Kriegel has argued for a similar distinction.

Lynne Baker and Luca Forgione have both emphasized the importance of first-person language in enabling the ability to consciously distinguish self from other and to conceive of one's self as one's self. With such "I" thoughts, a person does not need third-person noetic referential devices (a name or description) to know who she is, and thus, I suggest, conceive of herself autonoetically.

The "me, myself, and I" view of the self implies something contained within one's skin. But the pronoun "mine" also refers to what is known as the "extended self," an idea for which William James laid the foundation in this passage: "A man's Self is the sum total of all that he *can* call his, not only his body and his psychic powers, but his clothes and his house, his wife and children, his ancestors and friends, his reputation and works, his lands and horses, and yacht and bank-account. All these things give him the same emotions. If they wax and prosper, he feels triumphant; if they dwindle and die away, he feels cast down—not necessarily in the same degree for each thing, but in much the same way for all."

When you are aware of who you are, you are drawing from your self-schema, which is what underlies your self-concept. You self-schema includes your skills and abilities; your foibles; your social roles; your psychological attributes; your self-worth; how you look; how you feel and act; how your body responds in certain situations; what you expect your future to hold; how you feel about your family, friends, enemies, colleagues, acquaintances, and even your possessions, and un-possessed but desired things of either the natural or man-made world.

But your self-schema is not static. You change over time and are different in different contexts or situations. Damasio and others emphasize that the momentary self is a dynamic variant on a more stable core, or what Martin Conway refers to as the "working self." The working self reflects the self-schema components that are active at a particular moment. As with other schema, self-schema are nonconscious representations that are made active by pattern completion, and become the momentary working self.

The concept of "self" one has is dependent to a significant degree on the psychological conceptions that are embedded in his or her culture and its native language. In a classic paper, Hazel Markus and Shinobu Kitayama argued that a culture's view of the self greatly influences, and in some cases determines, the nature of how people experience themselves and the world around them. But ultimately, your experience of your culture is personal. Indeed, according to Nick Shea and colleagues, culture is made possible by the unique human cognitive system, which is capable of creating "suprapersonal" representations that can be shared with others. This is just another example of what autonoesis does for us.

I thus propose that noesis and autonoesis define much of the conceptual space that a theory of consciousness needs to cover. Tulving mainly used these terms to refer to the conscious experience of explicit memories in the absence of the remembered stimuli or events. But I think they can also be applied to immediate conscious experiences, which are also shaped by explicit memory—recall the earlier discussion about the role memory plays in top-down expectations and predictions that influence how we experience the world.

Because much research has been conducted on the brain mechanisms of semantic and episodic memory, Tulving's novel conception gives us leverage in understanding how important facets of consciousness might come about in the brain. In the next two chapters we look at the brain circuits underlying these kinds of memory and how such memories might contribute to conscious experiences.

*Chapter 57*

# PUTTING MEMORIES IN THEIR PLACES

In order to use explicit memory as a fulcrum for better understanding consciousness in the brain, we need to delve into its underlying brain circuits. We've seen that areas of the medial temporal lobe (perirhinal cortex, parahippocampal cortex, entorhinal cortex, and hippocampus) play a key role in explicit memory.

Information processing, as discussed earlier, was traditionally thought to change in the passage from secondary visual cortex to the medial temporal lobe. Semantic memory was believed to be mediated by connections from the item or object visual stream to the perirhinal cortex, and episodic memory by connections between the spatial/action (dorsal visual) stream to the parahippocampal cortex. The current view, discussed in part 11, is that visual areas are also involved in memory storage and medial temporal lobe areas are also involved in perception. The full set of connections involving the visual cortex and the medial temporal lobe are thus better thought of as perceptual-memory streams (or circuits) than separate sensory and memory circuits (see figure 48.4). The two perception-memory circuits converge in the hippocampus, allowing objects to be perceived in the context of complex scenes. This fusion of what and where information is in the hippocampus is a start toward episodic memory, but the temporal element is also needed. And indeed the hippocampus encodes time as well, allowing what-where-when representations.

The various medial temporal lobe areas are strongly connected with the medial prefrontal cortex: the perirhinal cortex of the semantic stream connects to the orbitofrontal cortex, the parahippocampal area of the episodic stream connects to the anterior cingulate cortex, and the hippocampus to the ventromedial area (figure 57.1). But the perirhinal and parahippocampal areas also connect with the dorsal lateral and ventral lateral prefrontal cortex areas (not shown in the figure), enabling explicit memory to contribute to executive goal processing.

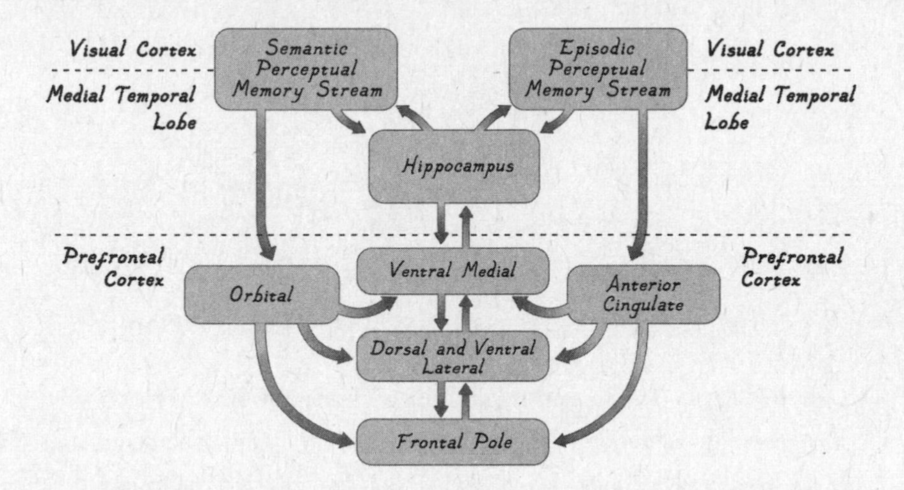


**Figure 57.1:** ***Perceptual-Memory Streams Converge in Prefrontal Cortex***

It has long been known that the medial temporal lobe is not the only location where semantic memories are stored. As mentioned in chapter 48, semantic and especially complex conceptual semantic memories are also stored in multimodal convergence zones in the neocortical temporal and parietal lobes (parietal temporal junction, superior temporal cortex, temporal pole). Additionally, language-processing areas store semantic labels (words) for objects and concepts.

Much attention is currently focused on the temporal pole, which is viewed as a neocortical semantic/conceptual hub that integrates across various unimodal and multimodal inputs to create general, abstract concepts and schema (figure 57.2). It does so by analyzing similarities and differences between individual items to form generalizations and inferences about what something is and is not, and makes possible item recognition from its appearance, sound, taste, touch, smell, and/or name. The temporal pole is one of the brain areas first affected in Alzheimer's disease and accounts for some of the early memory problems in this disorder. Similar to the prefrontal cortex, the temporal pole changed considerably in the transition from apes to humans.

Like the medial temporal lobe, the neocortical multimodal semantic/

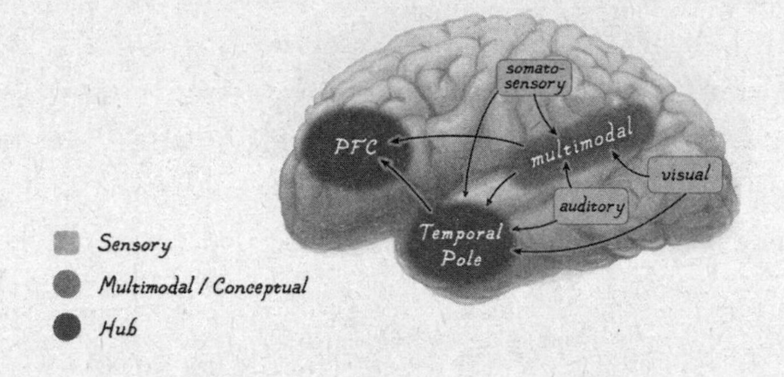


**Figure 57.2:** ***The Temporal Pole: A Semantic/Conceptual Hub***

conceptual networks also connect with areas of the prefrontal cortex. (These connections are shown in figure 58.1 in the next chapter.) Included are lateral prefrontal areas (dorsal lateral and ventral lateral), and the frontal pole; the latter, as noted, has the greatest capacity in the brain to create complex conceptual representations. The conceptual networks also connect with medial prefrontal cortical areas (orbital, ventral medial, anterior cingulate).

Through these diverse connections, the prefrontal cortex can integrate and use semantic and episodic memory in top-down control of information processing and deliberative behavior. For example, when an object is encountered, it is recognized as a meaningful entity by being referenced against perceptual and conceptual templates stored in both the medial temporal lobe and the neocortex. These representations form complex semantic schema and concepts that are then used by the prefrontal cortex in executive top-down control over information processing.

When episodic memories are based on multiple similar experiences that recur, or when the episodic memory of a single experience is retrieved repeatedly, they become "semanticized"—represented as a fact. For example, your experience of being at your workplace is a unique one when you are new at your job, but becomes "more of the same" the longer you are there, and can be represented semantically as simply "being at work," except when something novel happens there. When episodic memories become semanticized, they lose their dependence on the medial temporal lobe and come instead to be neocortically based. This is more likely to occur for older memories of recurring situations than for recent, novel ones.

Randy O'Reilly, Jay McClelland, and Bruce McNaughton proposed that the greater involvement of the neocortex in older memories results from the way that the hippocampus and neocortex learn: they suggest that the hippocampus is wired to learn rapidly, which is especially important in forming memories about unique facts or episodes, while neocortical areas slowly accumulate information across experiences, which is particularly useful in forming and adding information to concepts and/or schema.

An alternative view by Morris Moscovitch and Lynn Nadel is that the hippocampus remains involved even in old memories, but that copies of the memory come to be stored elsewhere in the process of learning categories and concepts. These nonhippocampal representations then support memory retrieval when the hippocampus is damaged.

While what-where-when representations are essential to episodic memory, they are not sufficient. Also needed, as we've seen, is the ability to engage in mental time travel, which requires a representation of one's self in time. Brain areas that have been implicated in self-processing (figure 57.3)

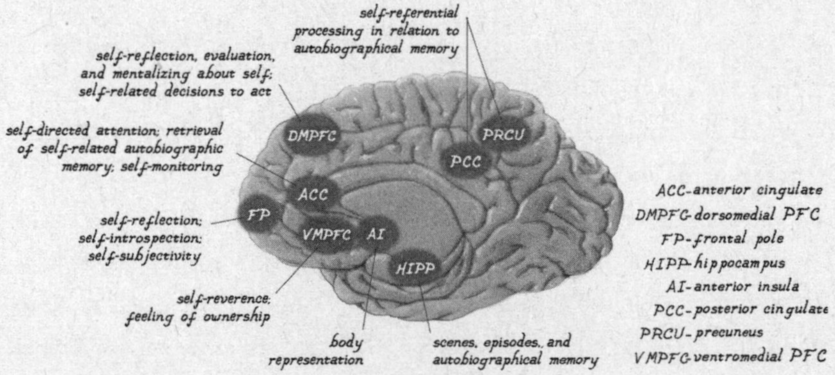

**Figure 57.3:** ***Self-Processing Areas in the Medial Cortex***

overlap, to some extent, with areas of the so-called default network of the brain, which is active during passive mental states, such as undirected "mind wandering." Included are the posterior cingulate/precuneus region, hippocampus, and anterior cingulate areas (figure 57.4). Randy Buckner and Daniel Carroll suggest that the reason the self and default networks overlap is that we are often thinking about our self when we are mind wandering.

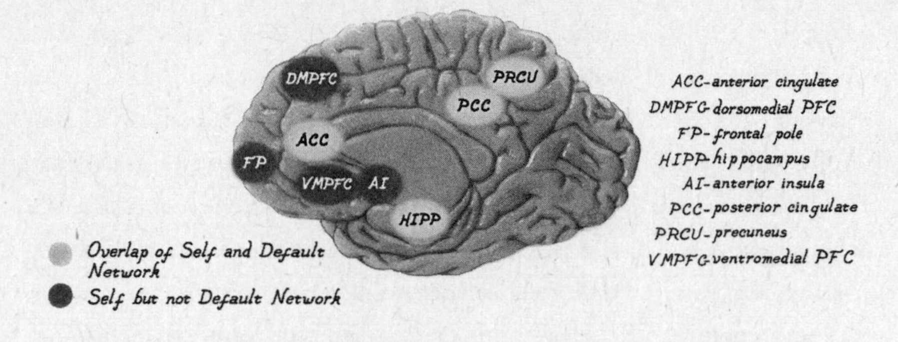


**Figure 57.4: *Default Mode Network Overlap with Self-Processing Areas***

*Chapter 58*

# HIGHER-ORDER AWARENESS THROUGH THE LENS OF MEMORY

Let's now turn to how research on memory helps give us a better, deeper understanding of consciousness, especially within the context of a higher-order account of consciousness. Specifically, I will develop a modified higher-order theory—*a multistate hierarchical model of consciousness.* To do this, I will start by using a standard view of higher-order perceptual awareness to help pinpoint brain circuits that might play a key role in consciousness, and then work backward from these to identify lower-order inputs to the network. This will reveal that while the higher-order network receives some sensory inputs, most of the inputs are from circuits involved in mnemonic and conceptual processing. Although I develop the model by emphasizing perceptual awareness here, I will later extend the model to account for emotional consciousness as well.

According to higher-order theory, a perception is consciously experienced when there is an appropriate higher-order representation of first-order, mainly sensory, information. Lau and Rosenthal's neural model, described in chapter 54, proposed that laboratory findings about conscious visual perception could be accounted for by interactions between the visual cortex and a prefrontal network involving dorsal and ventral lateral and polar prefrontal areas. I suggested some minor adjustments to enhance the explanatory power of this model, but also noted that even with these changes the model would not be able to account for complex real-life experiences involving, for example, memories and emotions. The addition of connections between the higher-order prefrontal network and memory circuits, I suggested, might be what is needed. It's time to show what I mean.*

Conscious awareness of complex visual stimuli and situations encountered in life is unlikely to simply involve prefrontal re-representation of pure sensory information processed in primary and secondary areas of the visual cortex. Secondary sensory areas, as noted earlier, store memory representations and use them in processing incoming sensory signals. Thus, at least some of the information sent to the dorsal and ventral lateral prefrontal cortices from sensory areas has likely been filtered by semantic memories stored in these areas.

The dorsal and ventral lateral prefrontal cortex regions also receive inputs from the multimodal convergence zone in the neocortical parietal and temporal lobes, areas that integrate inputs from multiple unimodal sensory processing areas, and form and store semantic memories, some of which are quite complex and conceptual (figure 58.1). Additionally, the dorsal and ventral lateral prefrontal areas also receive inputs from areas of the semantic and episodic memory streams in the medial temporal lobe (for example, perirhinal, parahippocampal, and hippocampal areas).

The dorsal and ventral lateral prefrontal cortices are also highly interconnected with other prefrontal areas, especially medial prefrontal areas. And these also receive inputs from the neocortical temporal and parietal multimodal convergence zones, and/or from the medial temporal lobe semantic and episodic memory networks.

The frontal pole, of course, is also part of the prefrontal higher-order network. It is richly connected with the dorsal and ventral lateral prefrontal areas, as well as with the multimodal convergence zones in the neocortical parietal and temporal lobes, and with medial temporal lobe areas.

All of the areas sending inputs to the higher-order awareness network can thus be said to be lower-order. This includes not just sensory processors, but also the various areas that are anatomically higher-order, but are

* The anatomical connections to be discussed in this chapter were all described in chapters 48, 49, and 50 of part 11 and in earlier chapters in this section. See the bibliographic entries related to those chapters for the relevant citations.

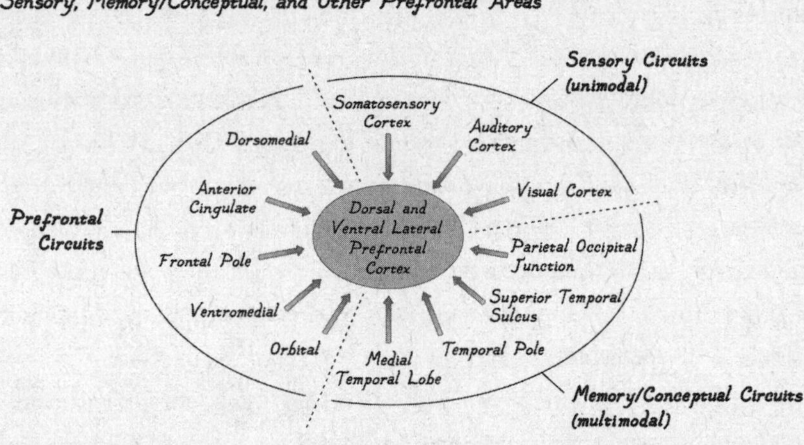


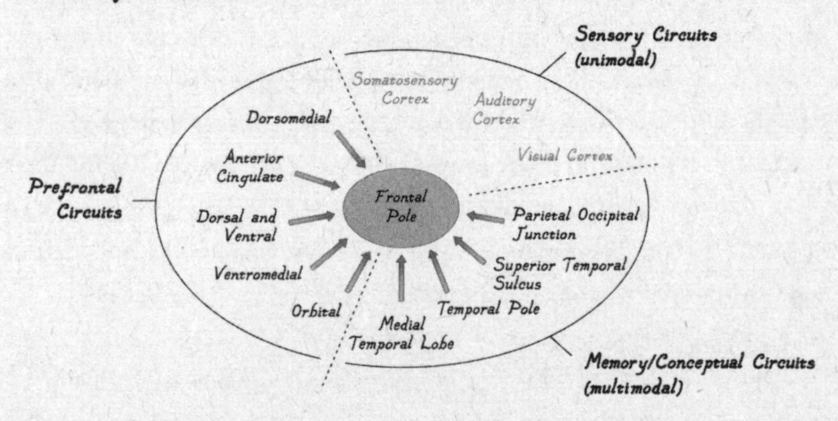


**Figure 58.1:** ***Diverse Lower-Order Inputs to Prefrontal Areas Implicated in Higher-Order Awareness***

lower-order with respect to the higher-order network. In other words, memory and conceptual processing regions are higher-order with respect to the visual cortex, and medial prefrontal areas are higher-order with respect to memory and conceptual processors, but all of these are lower-order with respect to the higher-order consciousness network.

Lower-order events processed by the higher-order prefrontal cortex are thus diverse and involve partly redundant representations with different degrees of abstraction that reflect their origin in multiple processing streams (sensory, memory, conceptual). The various streams also interact at multiple points in the bottom-up hierarchies, as do the various lateral and medial prefrontal areas that receive these diverse inputs. As a result, the focus of awareness (isolated sensory features of an object, the object itself, its name and categorical classification, and/or its role in a complex scene) may change as an episode unfolds over time. The momentary state may depend on which representation, or which combination of representations, is selected and moved into the higher-order network, and thus into the forefront of consciousness, depending on the long-term (frontal pole) and immediate (dorsal lateral) goals that are currently active in the network. Interactions across this entire set of connections likely determine what ultimately underlies a given conscious experience. And in any given situation, the balance between the relatively simpler (sensory) and relatively more complex (memory/conceptual) representations may differ.

Critics of higher-order theory, as we've seen, argue that prefrontal areas lack the kinds of fine-grain representations that could support the rich quality of perceptual experiences. But this criticism loses much of its relevance when perceptual consciousness is viewed, not simply as a sensory-dictated perceptual experience, but as a top-down construction based on combinations of sensory, memory, and conceptual representations that enable rich perception, via conception.

Earlier I noted that the classic idea is that "unconscious inferences" create mental models that shape what we experience. For instance, when you consciously notice an apple and verbally declare that you see it, it is not because the apple, in all its glory, is necessarily represented in your prefrontal higher-order network. You likely "see" a conceptual abstraction based in part on the lower-order sensory representation of the actual apple, but also on memories of the features of past apples that contribute to your schema/mental model of apples. The result is a representation that subjectively looks like what you *expect* an apple to look like, even if the pure sensory representation differs somewhat from this expectation, or even when it does not exist at all (recall the HOROR theory).

The pattern of inputs to the higher-order network is compatible with the multistate hierarchical higher-order model developed here, in which the network can draw upon a wide range of states, from pure sensory to mnemonic and conceptual, and many in between, in constructing schema/mental models that underlie our conscious experiences. It is tempting to speculate, given its inputs and role in highly conceptual processing, that the frontal pole might play an especially important role in conscious experiences based on schema/mental models.

One kind of mental model that we will discuss later in relation to our emotions is our self-schema. The frontal pole has, in fact, been implicated in conceptual self-awareness, the ability to use autobiographical memory in thinking about one's self, and thus in autonoetic consciousness. Helen Gallagher and Chris Frith proposed that this area also underlies our ability to use knowledge about one's self to understand the minds of others. Perhaps the frontal pole is the key to subjective states of all types, providing a kind of higher-order clearinghouse for subjective experiences, with the nature of the experience determined by the particular inputs being processed at any given moment. With inputs from other prefrontal regions involved in self-processing, it has access to autobiographical memories, including episodic and semantic autobiographical memories, and to thoughts about one's self, body states, and representations of one's ownership of brain and body states (figure 58.2).

Considerations of higher-order awareness have been hampered to some extent by the tendency of philosophers to put mental states in terms of propositional statements. For example, a common way to describe the higher-order representation that underlies perceptual awareness is "I am seeing an apple." Because this propositional statement takes the stance of someone who is using the personal pronoun "I," it leads to the assumption that the conscious self is part of the higher-order perceptual experience. As we have seen, there is a difference between being noetically conscious that

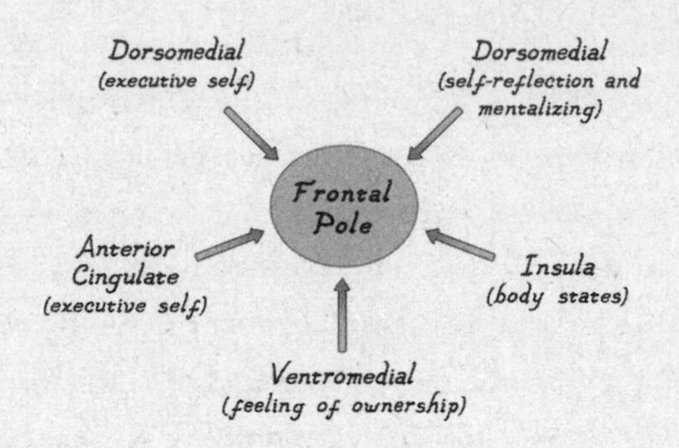


**Figure 58.2:** ***Self-Related Inputs to Frontal Pole from Other Prefrontal Areas***

an object is present (which is what the propositional statement "I see an apple" usually refers to in higher-order theory) and being autonoetically aware of one's self as a subject of the experience (an entity that knows that it is seeing an apple). The self as object may be a capacity we share with some other mammals, while the self as subject is likely ours alone, and is perhaps dependent on the polar frontal cortex.

In light of these speculations, it is important to revisit the criticism that damage to the prefrontal cortex does not consistently disrupt conscious experience. There are many points at which perceptual representations can be shaped and otherwise altered by memory/conceptual processing in the prefrontal network. Given this redundancy, brain damage might have to be quite extensive to take the system down completely. Second-string prefrontal teammates may fill in for injured collaborators for specific aspects of conscious experience. Indeed, anterior cingulate, orbital, ventral medial, insula, and other prefrontal areas outside the higher-order network share connections with one another, as well as with areas of the higher-order network. Cognitive processes that contribute to, but that do not themselves instantiate, phenomenal experiences must be distinguished from those that do. Until we know what each subregion contributes, it will be difficult to recognize and properly interpret exactly what about consciousness is impaired and spared following prefrontal damage.

While the scientific study of consciousness has made great progress by focusing on perception, we need to begin to take steps that will give us a fuller account of our experience. At a minimum, a theory of consciousness ultimately has to account for at least the following kinds of states: those that are about *fleeting and meaningless* perceptual events (events such as a flash of light or a brief sound); *lasting but still meaningless* perceptual events (an unfamiliar stimulus in isolation, such as a street sign in a foreign language); *meaningful* perceptions shaped by memory (recognition of a common object alone or in the context of a scene, such as an apple in a bowl with other fruits, or a song from its opening line); *absorbing* episodes of daily life (a conversation with a friend, an unpleasant encounter with a superior, the taste of a delicious dessert, an engaging piece of music or a painting, contemplation of one's own existence); *consuming* illness (chronic pain, pathological fear, anxiety, or depression); and perhaps many others.

We are still in the early days of research on consciousness. At this point no theory is safe from being eliminated by future research. Of all the theories currently on the table, I believe that global workspace and higher-order theory are most likely to account for the full range of conscious experiences we have, from relatively simple perceptual states to highly complex ones involving our memories and emotions. And of these two approaches, my money is on a higher-order account, especially one along the lines of the multistate hierarchical model proposed here.

## Part 14

# The Shallows

*Chapter 59*

# THE TRICKY PROBLEM OF OTHER MINDS

The picture of consciousness I've painted so far is extremely anthropocentric—human centered—as it should be. Human consciousness depends on unparalleled cognitive processes that are entwined with language and culture, and is enabled by circuits with unique properties. The evolutionary past of human consciousness, and autonoetic consciousness in particular, is, in my view, shallow rather than deep.

Autonoetic consciousness didn't arise out of the blue, but neither was it directly passed on to us from our animal ancestors. This idea won't sit well with many people, who don't have to look any further than their pets to find compelling evidence of consciousness in other animals. Behaviors enacted by our furry and feathered friends seem to signify emotions we are sure we see in them, and they certainly seem to reciprocate the love we give them.

Darwin, as discussed earlier, felt strongly about animal emotions: "A dog carrying a basket for his master exhibits in a high degree self-complacency or pride. There can, I think, be no doubt that a dog feels shame . . . and something like modesty when begging too often for food." But such intuitions based on analogy with human behavior are not scientific conclusions. The late-nineteenth-century comparative psychologist Herbert Spencer Jennings, who studied the behavior of protozoa, noted, "If *Amoeba* were a large animal, so as to come within the everyday experience of human beings, its behavior would at once call forth the attribution to it of states of pleasure and pain, of hunger, desire and the like, on precisely the same basis as we attribute these to the dog." Jennings pointed out that such attributions are useful to us, allowing us to "appreciate, foresee, and control" the behavior of other species. Larry Weiskrantz, whose book *Consciousness Lost and Found* alerted me to Jennings's remarks, noted that we similarly attribute humanlike qualities to objects in computer games, mechanical toys, and robots.

Every scientist is also a layperson, and brings everyday assumptions into the lab. We study psychological processes because we have psychological experiences, and want to know more about them. But we have to go beyond intuitions and analogies when being scientists.*

I have tried to the instill in my students respect for the importance, when being their scientific selves, of guarding against anthropomorphic assertions, and have urged them to avoid citing fear, for example, as the reason why rats behave the way they do in the presence of stimuli that portend harm. But our language is inherently anthropomorphic, and as a result our concepts and thoughts tend to lean in this direction as well.

J. S. Kennedy, writing in *The New Anthropomorphism*, notes: "Anthropomorphic thinking . . . is built into us. . . . It is dinned into us culturally from earliest childhood. It has presumably also been 'pre-programmed' into our hereditary make-up by natural selection, perhaps because it proved to be useful for predicting and controlling the behavior of animals." Indeed, the attribution of human mental states to animals is believed to have played a key role in their domestication, which occurred during the agricultural revolution in the late Neolithic period, around 3000 BC.

The fact is that research on animal behavior almost has to start from an anthropomorphic stance. We want to understand what's important to us. The reason so much psychological research in animals involves stimuli that elicit tissue irritation, eating, drinking, sex, and defense, is that we

* I was once at a dinner with a renowned physicist. He dismissed outright my contention we can't *really* know whether animals have conscious experiences. "Of course they do," he bellowed, and proceeded to talk at length about his dog. Now suppose someone from another field told him that quantum physics can't possibly be right because common sense suggests otherwise. He would certainly not give credence to such an intuition. But when it comes to animal consciousness, intuition, it seems, is all you need.

often have profound experiences in response to such stimuli. For example, we feel pain to tissue irritation, pleasure to certain foods and drinks or from sex, and fear when defending against danger.

Some psychologists and neuroscientists, including esteemed ones, claim that because such stimuli make us feel a certain way, if an animal behaves similarly to the way we do in their presence, the animal must feel what we do. For example, the highly regarded primatologist Frans de Waal expressed this sentiment when he wrote, "If closely related species act the same, the underlying mental processes are probably the same." Jane Goodall stated, as a matter of fact, that "animals feel pleasure and sadness, excitement and resentment, depression, fear and pain." She "knows" what animals experience because she has seen signifiers of these emotions in their behaviors.

If all we had to do to link consciousness to behavior was to observe behavior, we wouldn't need research. Observation is not sufficient. Much of what humans do as we make our way through daily life is done without explicit awareness. Even when we are aware that we behaved in a certain way, it does not necessarily mean that we consciously controlled the behavior. As we've seen, we learn a lot about ourselves by monitoring what we do. Larry Weiskrantz presciently noted that since consciousness is not always necessary for human perception and behavior, evidence that animals produce appropriate behavioral responses to visual stimuli does not qualify as evidence that they are conscious of what they are seeing.

Beyond a general tendency toward anthropomorphism, each scientist also comes to the lab with a set of personal biases and dispositions. The philosopher Bertrand Russell once noted, "All the animals that have been carefully observed have behaved so as to confirm the philosophy in which the observer believed before his observations began." Scientists should not attempt to prove that their personal biases, their preconceptions, are correct. Mahzarin Banaji and Anthony Greenwald explain that this is difficult because human nature is such that some of our strongest biases are not freely available to our conscious minds. If you aren't aware of a bias, you can't guard against it.

It is sometimes argued that an anthropomorphic stance toward animal consciousness should be adopted on moral grounds, and scientific evidence should be used to determine which animals deserve moral consideration. But Marian Dawkins eloquently argued that tying the moral issue to the science of animal consciousness weakens rather than strengthens the moral argument, because the question of which animals, other than humans, are conscious is difficult to determine scientifically. Quoting Thomas Huxley, she notes: "Do not pretend that conclusions are certain that are not demonstrated or demonstrable."

Some try to get around critiques of animal consciousness by calling upon "the problem of other minds." According to this philosophical argument, because we only have access to the outward behavior of other humans, the only person we really know is conscious is our self. And because we accept this kind of evidence in humans, we should do the same for other animals. But the "problem of other minds" is a hypothetical philosophical argument, not a scientifically based one. The fact is, as we've seen, no other animal has our kind of brain, especially our kind of prefrontal cortex, and the type of cognition it makes possible. And if the unique aspects of our brain and our cognition are key to our kind of consciousness, then our kind of consciousness should not simply be assumed in other animals on the basis of the other minds problem.

Regardless of whether you are studying humans or animals, rigorous methods should be used in the effort to obtain reproducible results and, ideally, unambiguous evidence. But a double standard exists in that different criteria are used in animal and human research on consciousness.

In human research, if you want to determine if a particular process depends on conscious or nonconscious mechanisms, you test both possibilities. For example, to confirm that people can respond to stimuli nonconsciously requires that you demonstrate that consciousness was not involved. As discussed earlier, critics are always proposing possible ways that a stimulus might have slipped through and thus have had some effect on consciousness. As a consequence, the criteria become more and more stringent.

Given this practice, it makes sense that a similar approach should be taken when trying to make the case for animal consciousness. In other words, the way to demonstrate consciousness in animals should be to compare conditions under which the behavior is best explained as being dependent on consciousness, and cannot be reasonably explained in terms of some nonconscious process.

But in animal research, this is not the standard. Most experiments are not designed to ascertain whether a particular behavior is consciously or nonconsciously controlled. Instead, they involve amassing more and more support for the intuition that consciousness was involved. Given his status as a renowned scientist early in his career, and later as a philosopher, Nick Humphrey once made a rather strange suggestion: "Away with critical standards, tight measurements and definitions. If an anthropomorphic explanation *feels* right, try it, and see; if it doesn't feel right, try it anyway."

Each species manifests its own exquisite form of adaptation that made it possible for it to survive in its own unique way. Some animals may have the ability to be conscious of what they are sensing and doing, and even feeling, in given situations. But can we really ever know this with confidence scientifically without lowering the bar for what counts as a genuine conscious experience?

It's hard for us to imagine complicated behaviors being carried out nonconsciously, since we are usually conscious when we do such things ourselves. But this should not be the basis for the conclusion that consciousness was involved in a given behavior in another organism. The scientific question in an experiment is not whether animals might be conscious in some general sense, but instead whether consciousness specifically accounts for the behavior that was studied. If this is not tested, the statement that consciousness was involved is not warranted scientifically.

Frans de Waal uses the term "anthropodeniers" for those who suggest caution in making claims about emotions and other conscious states in animals. Is he drawing a comparison with distinctly anti-science "climate-change deniers"? This would be odd, given that the so-called anthropodeniers are the ones demanding more rigorous scientific standards. But more to the point, no one is actually denying anything. It's not about dismissing animal awareness outright but instead about considering what we can and can't say with the data we collect.

Standards, measurements, and definitions are not optional. They are what elevate science above commonsense intuitions based on analogies with human behavior. An excellent example comes from Alexandra Horowitz's studies of dogs. She has shown that pet owners' intuitions about the conscious thoughts and feelings of their canine friends are typically wrong, and argues that the behaviors are more accurately accounted for by underlying cognitive processes that do not require consciousness.

Why don't more animal researchers take this approach? I think part of the reason is that research on human and animal consciousness has proceeded in relative isolation, and animal researchers don't necessarily know about standards that are used to assess distinctions between conscious and nonconscious processing in humans. Perhaps some know about these approaches but are frustrated, because when the scientific steps required to evaluate nonconscious versus conscious accounts are taken, the conscious explanation is often not the best explanation of the result. On the other hand, as noted above, some researchers are committed to the existence of animal consciousness for nonscientific (moral/ethical) reasons. Animal ethics is an important topic with consequential implications for research, but it should not be conflated with what counts as criteria for demonstrating consciousness in scientific studies.

Why is it so hard to demonstrate consciousness in animals? Previously we saw that when people are conscious of a stimulus, they can report on their knowledge either verbally (by saying "I see an apple") or nonverbally (by manually picking the apple from a bowl that also has a pear and orange in it). But when responding to a subliminal (unconsciously processed) stimulus, humans can only respond nonverbally; they can't give a verbal report of something that they didn't see (figure 59.1). The fact that animals can only respond nonverbally means there is no other response to help distinguish conscious from nonconscious processes. The lack of language in nonhuman animals makes it very difficult to scientifically know what, if anything, is on their minds.

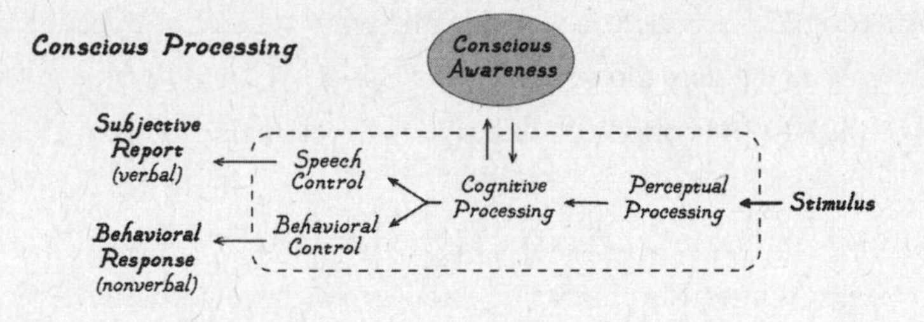


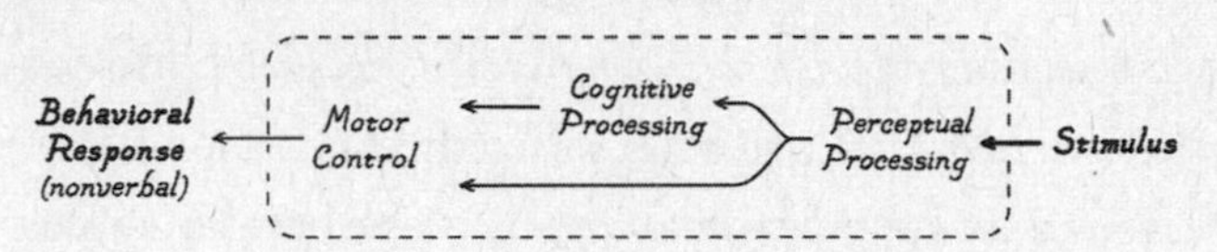


**Figure 59.1:** ***Conscious Processing Can Be Reponded to Verbally or Nonverbally, but Nonconscious Processing Can Only Be Responded to Nonverbally***

Verbal reports are not without controversy. Some researchers have a reflexive response against the use of verbal reports as scientific data, arguing that they are notoriously unreliable. But this conclusion is sometimes based on a misunderstanding. The fact is, verbal reports are actually fairly reliable indicators of what a subject is experiencing in real time, as people tend to know what they are conscious of, especially if probed in the moment. Such reports become less reliable over time, even over short time spans, since memory is dynamic—it changes by the mere passage of time, or by the accumulation of new experiences. For example, an eye witness who reads or hears about a crime in the news may later unwittingly report information from the news story.

Another limit of verbal reports reflects the fact that experiences can be fleeting and unreportable—how often, for example, have you suddenly lost a thought? People may also not be truthful about what they actually experience. Dishonesty is more likely to be a factor when questions are of a personal nature—such as "How many sexual partners have you had?"—than in a typical situation in a perceptual study where the question might be "How many letters did you see on the screen?" In general, a verbal report is a useful indication of what one is experiencing now in an experiment.

There is one exception. Daniel Kahneman, Tim Wilson, Richard Nisbett, Gerald Clore, Norman Maier, and others have pointed out that the most significant limit of verbal reports emerges when people are asked about their motivations—about why they did what they did. When behaviors are controlled nonconsciously, they are not necessarily consciously accessible and cannot be reported reliably. Yet, as in split-brain patients, people generate narratives to account for their behavior.

Because of the limits of verbal reports, some have attempted to develop methods that assess consciousness using more "objective" measures, such as confidence ratings about decisions. For example, rather than simply asking participants to say what they see, they are also asked how confident they are in their report. Summarizing this approach, David Rosenthal notes: "Confidence ratings work reasonably well in un-problematic cases for assessing subjective awareness. Still, when confidence ratings diverge from subjective report, it is reasonable to let subjective report override confidence ratings." He continues: "Sincere report is a reliable indicator of whether a psychological state is conscious because sincere report reveals whether the subject is aware of being in that state. . . . And being aware of a psychological state is in turn pivotal for consciousness because if an individual is in some psychological state but in no way aware of being in it, the only credible explanation of that lack of awareness is that the state is not conscious." Further, leading researchers in this area, like Hakwan Lau and Steve Fleming, point out that these "objective" measures are useful because they may tap into some of the same higher-order neural mechanisms as consciousness, not because they measure consciousness itself.

The bottom line is that the best way to probe whether a state is conscious in a scientific experiment remains simply asking the subject what, if anything, was experienced. Short of that, the most you can hope to do is to use nonverbal behavioral guesstimates to try to creep up on consciousness. And that is always the case when the subject in question lacks language—in other words, is not a human.

*Chapter 60*

# CREEPING UP ON CONSCIOUSNESS

To illustrate the difficulties of performing scientific studies of consciousness in animals, I will consider some examples of the ways scientists have used behavioral proxies to, as Jeffrey Gray put it, "creep up on consciousness."

One of the most popular behavioral proxies of consciousness in animals is Gordon Gallup's mirror recognition test, which relies on a behavioral response that indicates that an organism recognizes some unexpected change in its own appearance in a mirror reflection (figure 60.1). For example, if a red dot is put on a chimp's face, and the chimp stares and touches it, Gallup assumes that it has recognized itself, and thus is self-aware. Researchers have used versions of the mirror recognition test to claim self-awareness in dolphins, elephants, birds, and other creatures. Gallup insists that animals other than apes pass only if the criteria are loosened, and that his test, when properly used, only shows self-awareness in humans and apes.

**Figure 60.1:** ***The Mirror Self-Recognition Test***

Others maintain that Gallup's own version of the test does not qualify as a means to demonstrate conscious self-awareness, even in humans and apes. Cecilia Heyes and Daniel Povinelli, for example, both argue that the results of the mirror test can be accounted for by invoking nonconscious behavioral strategies involving simple associative learning. Johannes Brandl and Thomas Suddendorf propose explanations that are cognitively leaner than Gallup's but richer than what Heyes and Povinelli allow. Still others, like Diana Reiss, avoid the minefield of animal consciousness altogether by emphasizing the importance of the findings for showing how sophisticated animal cognition can be, in her case involving dolphins, independent of whether self-awareness is involved.

Other studies have sought to determine whether animals are capable of episodic memory. To do this, they are tested on whether they can form and use representations that involve combinations of two, or sometimes all three, of the what-where-when elements that make up an episodic experience in humans. Apes, monkeys, rodents, and some birds have passed such tests. Especially impressive are Nicola Clayton's studies in birds. She has shown that shrub jays hide worms in different places, and then eat the oldest ones first, suggesting that they have what (worm), where (location), and when (oldest first) memories. One issue in such research is whether the animals actually formed unified what-where-when memories, as opposed to simply using multiple individual semantic memories, or whether they use simple mere associations. But the most important question, even if they do have a unified what-where-when memory, is whether they are conscious of it, and if so, if they have an autonoetic conscious awareness of themselves as part of the memory episode.

Elisabeth Murray, Steve Wise, and Kim Graham argue that claims about subjective episodic experience in studies of memory in animals are, by necessity, based on "unverified (and often unverifiable) assumptions." And precisely because of the difficulty in drawing conclusions about subjective experience in animals, Clayton and some other researchers are careful to refer to their findings from animals as showing "episodic-like" memories, recognizing that the behavioral results are consistent with some features of human episodic memory, but do not fully demonstrate it, and certainly do not demonstrate autonoetic consciousness.

Another proxy that some have used involves testing internal representations of time. As we've seen, Tulving argued that mental time travel—the ability to project one's self into their episodic past, or to imagine one's self in their episodic future—is a key feature of human autonoetic consciousness. Findings showing prospective (future-oriented) cognition in some species of animals have been used to argue that they may be capable of such temporal self-projections. But Cecilia Heyes, Sara Shettleworth, Michael Corballis, Thomas Suddendorf, and others, argue that the demonstration of future-oriented behavior does not necessarily qualify as mental time travel, since the latter requires the ability to actually experience one's self as being in the past or future. These critics maintain that leaner interpretations, based on nonconscious cognitive processes or noncognitive behavioral strategies, are sufficient to account for the data.

Clayton and Anthony Dickinson give a very concise, reasonable assessment of the work on episodic memory and mental time travel in animals: "In the absence of language, there is no way of knowing whether the jays' ability to plan for future breakfasts reflects episodic future thinking, in which the jay projects itself into tomorrow morning's situation, or semantic future thinking, in which the jay acts prospectively but without personal mental time travel into the future."

"Theory of mind" is a special kind of cognition in which one has thoughts about other people's mental states (figure 60.2). In social interactions we use knowledge about how our own mind works to read others' minds—to attribute thoughts, beliefs, and feelings to them, and predict their behavior. Uta Frith and Francesca Happé note that this capacity is impaired in autism, and suggest that because afflicted people lack the ability to mentally represent their own thoughts and feelings, they cannot imagine these in others. Frith and Happé suggest that people with autism tend to be mired in first-order representations about the world, and are unable to create the kinds of higher-order representations required to be self-aware.

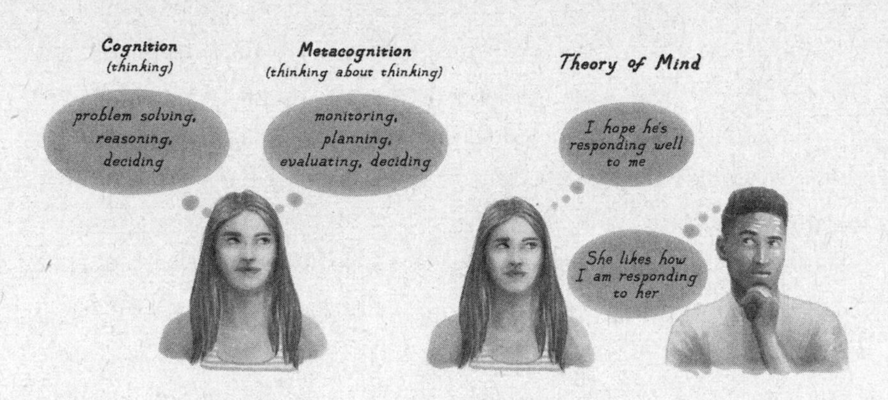


**Figure 60.2:** ***Cognition, Metacognition, and Theory of Mind***

This raises the question of whether other animals can engage in such "mind reading" when they interact with one another, and if so, whether consciousness is involved. As with other proxies, animals pass the behavioral tests for mind reading, but the evidence does not demonstrate that they are conscious of what they are doing. Heyes argues that animals don't need a theory of mental consciousness to predict the behavior of other animals any more than they need a theory of toxicology when choosing not to eat rotting food—visual stimuli and odors are sufficient. Clayton and Dickinson's cautions about using behavior to attribute conscious mental states in studies of episodic memory and time estimation are apt here as well.

David Premack, who started the animal theory-of-mind field decades ago, concluded, after years of disappointing results, that if animals are not "mentalists" (conscious mind readers), they are "behaviorists" (stimulus-response machines). But Premack postulated this distinction before nonconscious cognition was such a well-established part of cognitive science. Nonconscious cognition constitutes a middle ground between full-blown conscious mind reading and stimulus-response behaviors. Indeed, Chris Baker, Rebecca Saxe, and Joshua Tenenbaum note that theory-of-mind inferences about other minds are typically effortless, automatic, and unconscious, even in humans.

I agree with those who insist on the careful consideration of alternatives before concluding that conscious states account for animal behavior. Their efforts are helping to hold back the rising tide of anthropomorphism in science. For my part, I assume brains were nonconscious long before they were conscious. So my default position is that behavior is controlled nonconsciously until proven otherwise. As consciousness researchers Stan Dehaene, Hakwan Lau, and Sid Kouider recently noted, the human brain is the only physical system that unequivocally possesses consciousness.

*Chapter 61*

# KINDS OF MINDS*

Where does all this leave us? The first thing to note is that just because it's difficult to demonstrate consciousness in animals does not mean they are mindless. I don't know of any serious scientist today who claims that mammals and birds lack minds, if by "mind" we mean the ability to think, plan, and remember. But that is different than the ability to be conscious of one's own thoughts, plans, and memories. The fact that mammals and birds do have minds does not mean that they have the kind of mind that humans have, one that is capable of language and reflective self-awareness, of visiting one's own past and envisioning one's self in different possible future scenarios.

I thus concur with Tulving that autonoetic consciousness is a unique human feature. However, others, such as Christophe Menant and Michael Beran and colleagues, have suggested that a form of nonverbal autonoetic consciousness may exist in apes. I recently watched Fred Rogers (Mister Rogers) interact with the gorilla Koko in the film *Won't You Be My Neighbor?* Even an agnostic like me found it hard (almost impossible) not to intuit a very sophisticated conscious mind was at work in this animal. But that was me the human being, not me the scientist. Upon reflection, the scientist asked, How could I test that intuition? And, as I've stressed over and over again in these pages, doing so isn't easy.

But nonconscious cognition and reflective self-awareness are not the only two psychological options—noetic consciousness, based on semantic

* This title is borrowed from Daniel Dennett's 1996 book of the same title.

knowledge, falls in between them. In humans, noetically conscious states are often tied to our verbal semantic memories. Nevertheless, we also have nonverbal semantic memories. Even though we may verbalize a visual concept (tree, bird, river, food, mate), we can also use nonverbal conceptions to guide behavior, just as other animals do. Perhaps a nonverbal form of noetic awareness might exist in nonhuman primates, and possibly other mammals, and maybe even in birds. (As noted in chapter 60, Nicola Clayton and Anthony Dickinson proposed something along these lines.) With this kind of consciousness, an animal could be aware of being in the presence of danger, food, or mates, and perhaps be aware of memories about objects and situations; it might have a sense of familiarity in recurring situations, and, as Antonio Damasio has suggested, possibly even a sense of self versus other based on memories of body sensations (self as object) but without the knowledge that the experience belonged to it, and it alone (self as subject).

Research has been conducted on what might be thought of as the neural basis of noetic consciousness in animals, though it's not usually viewed in this way. For example, Nikos Logothetis and colleagues, studying monkeys' responses to visual stimuli, argue that they have found possible neural correlates of conscious perception. Specifically, they found that areas of the ventral perceptual/memory stream and the prefrontal cortex are active when monkeys respond behaviorally to stimuli under some conditions, but that under other conditions only the visual cortex is active. As we have seen, the combination of visual and prefrontal activity is what is found when humans can report on their subjective visual experiences. The findings are thus consistent with noetic visual consciousness.

However, we have to remember that even in humans, prefrontal activity is not a surefire way to implicate consciousness in behavior, since prefrontal circuits process information both consciously and nonconsciously. Consciousness needs to be measured, not assumed from the observation of activity in a brain area, and that is hard to do in nonhuman animals.

Other recent studies have found that prefrontal networks ignite global brain states in monkeys similar to what has been proposed by the global neuronal workspace theory of human consciousness. While these findings are also compatible with a noetic form of consciousness resulting from global broadcasting, the findings are equally compatible with a nonconscious interpretation since, as mentioned previously, the global neuronal workspace hypothesis, even in humans, does not clearly distinguish cognitive states that are conscious from those that are nonconscious.

The distinction between autonoetic and noetic consciousness makes it much easier to consider consciousness in other animals scientifically. Yet, as scientists, we have to also recognize that evidence that animals are able to use nonverbal semantic knowledge to solve complex, even very complex, problems is different from evidence for noetic consciousness. Without the ability for verbal report, it's hard to rule out nonconscious explanations, and do more than creep up on consciousness. So the idea of noetic consciousness still doesn't solve the problem of how to demonstrate animal consciousness unambiguously. But it does allow us to at least conceive of how animals could be conscious of events and their meaning as they transpire, and even have an awareness of their self as an object. For example, as noted earlier, it might allow the animal to be aware of the presence of nutritious versus poisonous food, or of friend versus foe, or of a potential mate, and also of body sensations, but without also having the more elaborate capacities underlying the ability to be reflectively aware of themselves as a participant in such states in the present or past, or in the imagined future.

To reiterate a point I made in the prologue, just because animals may not suffer the way we do does not mean that they do not experience some form of distress and discomfort, and suffer from body injury or illness in their own way. My position should therefore in no way be used as a rationale for torture, abuse, or other mistreatment of animals. With good reason, we have societal standards for the humane treatment of animals. Any activity that falls outside of what is allowed is, as it should be, a criminal offense.

Regardless of the status of animal consciousness, comparative research across species has taught us much about animal cognition, and thus about the roots of human cognition. Such research has shown that birds have cognitive problem-solving capacities that rival those of monkeys and apes, despite not being in the evolutionary history of primates, and having small brains. This likely occurred by way of so-called convergent or parallel evolution in which some capacity, such as prospective cognition, evolved in different groups because it was useful in their lives. For example, Michael Beran suggests that scarcity of food could have made it advantageous for birds to have the ability to plan for the future when storing worms or other items and when deciding when to eat what, while forward cognition may have evolved in nonhuman primates to solve problems related to planning travel routes or inhibiting behaviors that might draw attention from dominant males in social groups. Evolutionary history indeed offers many examples of similar traits resulting from different selective pressures, such as the opposable thumb of opossums, pandas, and primates.

Maybe if Wittgenstein's talking lion actually existed, *and* we understood what he was saying, the question of animal consciousness could finally be addressed. While this is an appealing idea, the reality is not that simple. As noted, there have been many attempts to demonstrate language abilities in nonhuman animals, including apes, monkeys, dolphins, and birds, among others. While some researchers claim success, others argue that language per se is not needed to explain the findings, since learned behavioral strategies or other nonconscious explanations are possible in all cases.

So far, the results of animal communication studies have not done for animal researchers what verbal reporting accomplishes for human researchers—namely, to serve as the means of distinguishing behaviors based on conscious versus nonconscious processes. This certainly doesn't mean that consciousness exists only in humans. But it does mean that care should be taken when making scientific claims about consciousness in nonverbal animals.

David Rosenthal argues that consciousness requires the capacity to have higher-order thoughts about mental states, and this, in turn, requires language. As a result, creatures without language cannot be in mental states that are conscious, at least not the way we are.

Ultimately, the debates about animal consciousness are connected to Darwin's proclamations about the mental life of our ancestors and relatives. Scientists, according to Clive Wynne, have struggled for well over a century with the contradiction between Darwin's ideas about mental continuity between humans and other mammals, on the one hand, and the obvious cognitive gap that exists, on the other. Darwin managed to remove the "scale of nature" in biology, but his anthropomorphism helped perpetuate it in psychology, compelling so many to search for and uncritically view behavioral continuity in terms of mental equivalence—all based on behavioral observations alone.

Wynne rightly notes that psychologists should follow the lead of contemporary biologists and give more attention to species differences, and in particular recognize that not all such differences are the same. Some are due to natural processes by which species evolve from common ancestors (such as cognitive changes in the evolution of humans from primate ancestors), but others are due to the evolution of similar capacities in species that do not share common ancestors (such as the capacity for vocal communication in birds and humans). Wynne uses as an example the fact that much work was done on apes in an attempt to identify hidden language abilities, under the assumption that since they are our closest animal relatives, there must be traces of human language abilities in them. This was a reasonable assumption, but the effort failed. Though language may well be a novel human trait, birdsong learning is a better animal model of human language acquisition than what was tried in apes. Why? Not because birds' vocal capacities and our own are based on shared common ancestry, but because the behavioral and neural similarities between birdsong learning and human speech acquisition reflect similar, but independently evolved, solutions to the problem of using sounds to communicate. And if this capacity, or key components of it, did not arrive until humans emerged, studies of apes would be limited in what they could reveal.

Because of the difficulty in scientifically measuring consciousness in animals, we may never be able to know for certain whether any other creatures have conscious experiences, and if they do, whether they are like those we have. But in a sense this is not the most important question for cross-species comparisons. We should be focused less on consciousness itself than on cognitive and behavioral capacities that are clearly shared with and measurable in other animals. Some of these have certainly contributed to the evolution of our kind of consciousness, even if they do not make other animals conscious the way we are.

The old idea, handed down from Descartes, that humans have conscious minds and animals are reflexive automatons, has long been discarded. Both animals and humans have cognitive processes that control behavior nonconsciously. The way to narrow the gap between humans and animals is thus not to make untestable assumptions about animal consciousness. Instead we should explore the cognitive underpinnings of behaviors across species, identifying similarities and differences and determining whether comparable neural mechanisms are responsible for them. The good news is that this is currently a vibrant research strategy at present. But for it to have the greatest impact, the lingering influence of late-nineteenth-century anecdotal mental anthropomorphism, and the extravagant psychological claims this influence continues to inspire, need to be exorcised. I have done my best to achieve this here.

# Part 15

# Emotional Subjectivity

*Chapter 62*

# THE SLIPPERY SLOPES OF EMOTIONAL SEMANTICS

Our emotions are the conscious experiences we care most about. But they have been particularly difficult to study scientifically, in large part, I believe, because of the way scientists have relied on everyday language to explain "emotion" as a scientific construct. I maintain that the words commonly used to describe emotions do have a place in science, but that they have been utilized inconsistently, and often inappropriately.

Jack Block pointed out some years ago, "Psychologists have tended to be sloppy with words." Take the word *fear,* for example. Like most terms describing emotions, it is used in reference to conscious experiences (the feeling of fear), to physical responses (freezing or fleeing; changes in circulatory, respiratory, and endocrine systems; or brain arousal), to the motivation to perform instrumental responses like avoidance behavior (fear drive), to cognitive evaluations (appraisals), and to all of the above together. Given this multiplicity of referents at multiple conceptual levels, it is not surprising that when an emotion word like *fear* appears in a scientific context, there can be some confusion about what precisely is meant.

Scientists often approach their work from their own personal vantage point, based on intuitions about some phenomenon they are interested in understanding. They are then supposed to leave the intuitive part behind and design tests to get to the underlying nature of the phenomenon. But when the scientific topic is a psychological one, special problems arise, as it's much harder to leave intuition out of the process. As George Mandler and William Kessen pointed out in their 1959 book, *The Language of Psy-*

*chology*, "Atoms do not study atoms, stars do not investigate planets. . . . The fact that man studies himself and that he has archaic notions which persist in the daily behavior of all men puts a major stumbling block in the path of scientific psychology."

One of those archaic notions is the idea that behavioral and physiological responses that occur in connection with our emotional feelings are actually caused by those feelings. Darwin, as we've seen, ascribed to this long-held bit of folk wisdom. Indeed, such intuitions are perfectly natural to humans, including most emotion scientists, as people do often feel afraid when fleeing or freezing in response to danger. Such intuitions can become assumptions. For example, the mental state of fear and its behavioral ambassadors are so entwined it seems they must be bundled together in the brain—otherwise, why would they appear together with such consistency? And assumptions can become dogma—brain areas that control the behavioral and physiological responses are also home to the feelings. Although every scientist knows that correlation does not equal causation, some things seem so obvious that a causal connection is simply presumed, and becomes a scientific truism, a fact, a dogma, and goes unquestioned.

William James in the heyday of late-nineteenth-century mentalism raised a red flag, questioning whether fear is the reason we run from a bear. Because James was more interested in the question of how the felt experience of fear and other emotions come about, he didn't have much to say about why we behave the way we do, except that feelings aren't the cause. The behaviorists took James's concerns a step further, eliminating inner states altogether as a relevant topic in psychology. But they retained inner state, and especially *mental state,* words as monikers to describe relations between stimuli and behavior. *Fear,* for example, was used to describe the stimulus-response relationship that characterizes dangerous situations. But after the demise of behaviorism, researchers were less constrained about proposing that subjective experience underlies behavioral responses to danger, both in animals and humans.

Research done by Peter Lang in the late 1960s should have been a sign that maybe James was correct in questioning commonsense assumptions about the role of fear and other emotions in behavior. His studies, which have since been replicated several times, showed that the correlation between the subjective experience of fear and concurrent behavioral and physiological responses is weaker than people typically assume from their own experiences.

The reason that people feel the connection is stronger than it is may be due to confirmation bias. We inherit a significant body of beliefs from the folk psychological assumptions of our culture. And once such a belief is in place, it serves as an unquestioned underpinning of our intuitions, which guide our actions. Instances that are inconsistent with the belief are then discounted and ignored.

Confirmation bias not only affects laypeople but also scientists. For example, the brain area most commonly assumed to be responsible for fear is the amygdala. The connection between the amygdala and fear was first made in the 1950s. But the idea that the amygdala is the brain's "fear center" began to heat up as a result of research done by Bruce Kapp, Michael Davis, and me, using Pavlovian fear conditioning in the 1980s. With this procedure, after being paired with a mild electric shock, a meaningless stimulus such as a tone comes to elicit defensive behaviors (such as freezing) and physiological adjustments (such as changes in heart rate, blood pressure, and hormone levels). The research we did demonstrated that the amygdala was an essential part of the brain circuitry that controls the behavioral and physiological responses elicited by the conditioned threat. And since we were studying "fear" conditioning, the idea naturally arose that a state of fear is what gets conditioned, and that the amygdala is thus a fear center (figure 62.1).

This intuitively appealing idea stimulated a tremendous amount of research, and also captured the interest of the lay public, in part through my 1996 book, *The Emotional Brain*. Today the amygdala fear-center notion is not only a scientific doctrine but also a cultural meme that appears routinely, and without question, in books, magazines, movies, songs, cartoons, and other media. Nevertheless, I think it is wrong. And since my research and writings contributed to this mischaracterization, I have some explaining to do.

In 1620, Francis Bacon wrote: "Scientists should be vigilant . . . and especially should guard against tacitly granting reality to things simply because we have words for them." In other words, when we name things, we reify them, endowing them with the properties implied by the name we give them. In the 1950s, Melvin Marx warned that the use of subjective-state names as labels for nonsubjective states that control behavior runs the risk of infecting behavior with subjective-state properties implied by the name. Indeed, when we name behaviors and the circuits that control them using emotion words like *fear,* the behaviors and circuits acquire the emotional implications of the name.

I was not aware of Bacon's or Marx's writings when I wrote *The Emotional Brain* but had, in a way, come to a similar realization. I started studying emotion in animals because of the findings Mike Gazzaniga and I obtained from split-brain patients suggesting that behaviors controlled nonconsciously prompt the generation of a narrative that contributes to the resulting conscious experience. I felt that emotional behaviors were prime examples of this, and that the circuits that control these might be readily studied in animals, even if conscious experience had to be studied in humans. And this is what I did using the fear conditioning procedure, which took me to the amygdala.

From the earliest days of my animal research I maintained that the amygdala was responsible for the control of so-called fear responses, but not for generating the conscious feeling of fear. To conceptualize this difference semantically, I borrowed the explicit-implicit distinction that was emerging in memory research. Specifically, I proposed that the amygdala is responsible for nonconscious or implicit fear in its control over responses. Explicit conscious fear, on the other hand, I argued, emerges from cortical cognitive circuits that are responsible for other conscious experiences. The amygdala thus contributes to the conscious feeling of fear, but only indirectly, and is not itself responsible for it (see figure 62.1).

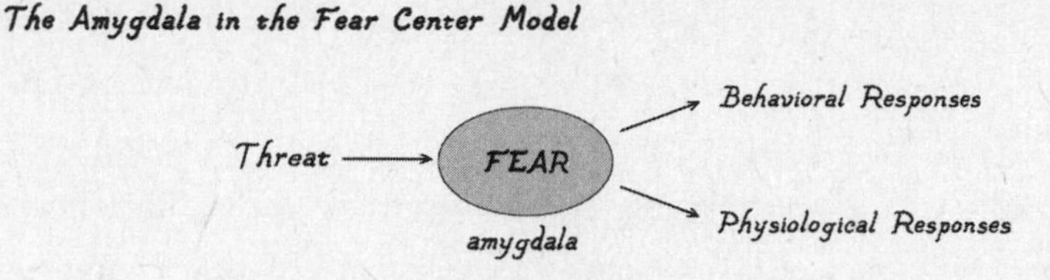


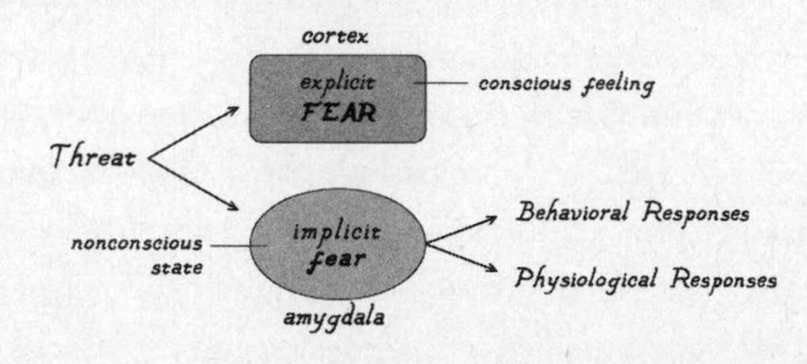


**Figure 62.1:** ***Two Views of the Amygdala's Role in Fear***

At the time this seemed like a useful way to characterize the mechanism underlying fear. It was consistent with research around the same time showing that when brief, masked pictures of threats are presented to people subliminally, the amygdala is activated and body responses are elicited, but without the person knowing the stimulus is present and without feeling fear. If the responses could indeed be elicited without the person feeling fear, it seemed that fear itself could not be responsible for the responses. Work by Adam Anderson and Elizabeth Phelps in 2002 indeed showed that conscious emotional experiences could still occur in people who suffered amygdala damage, a finding confirmed a decade later by Justin Feinstein and colleagues.

Nevertheless, many people (including laypersons and scientists) were not aware or chose to ignore the distinction between implicit and explicit fear. As a result, fear conditioning research was often assumed to implicate the amygdala in the genesis of fearful feelings. So instead of being part of an implicit-fear circuit, the amygdala was simply assumed to be a fear center. And without the qualifying adjective (i.e., *implicit*), fear was assumed to mean *conscious* fear.

I have to take some of the blame for the confusion, as I was sometimes sloppy with wording in my writings.* Because I contributed to the confusion, I hope that I can contribute to the solution.

It may seem obvious that scientists should be as careful in how we interpret and discuss our findings as we are in collecting and analyzing data. But sometimes the obvious is not apparent when you are in the thick of things. With hindsight on my side, I have been advocating for my field to be more precise in discussing fear and other emotional states. I believe we should clean up the language of fear, since the way we talk about our work has profound effects on the way we think about it and how we proceed in our research, and also on how it is applied clinically.

To illustrate the problems that can result from vague language, consider the elegant studies of flies conducted by the neurobiologist David Anderson. When flies encounter danger, they stop moving. By way of analogy with rodent work, Anderson calls this response "freezing." Though flies lack an amygdala, they have their own threat-detection circuit that controls their kind of freezing. He has shown that the genes underlying such behaviors are similar in flies and mammals, and an intriguing possibility is that these evolutionarily distant groups inherited circuit-constructing genes from their common ancestor, the PDA, which lived roughly 600 million years ago. The genetic similarity might also be the result of parallel evolution. In either case, Anderson has done impressive research. But I believe that he interprets his findings counterproductively, claiming that an emotion state, "possibly analogous to fear in mammals," occurs between the threat and the freezing behavior in flies, and that by studying flies we can learn important things about human emotions.

Some scientists today who talk about fear, hunger, and pleasure in animals are careful to qualify their use of this terminology and explain that they do not mean conscious feelings, but rather brain states that control

* Although I emphasized the amygdala as an implicit fear processor in *The Emotional Brain,* in the same book I described it as a "hub in the wheel of fear," with no qualification about implicit versus explicit meanings. And in my scientific papers I often talked about fear conditioning without mention of the distinction between conscious and nonconscious fear. In 2006, I wrote and recorded a song, "All in a Nut," for my band The Amygdaloids, that implied that the amygdala is the source of fear, again ignoring the distinction between implicit and explicit.

behavior (essentially what I was trying to convey with the "implicit fear" idea). But the fact is that most scientists, even those who note that they are not talking about fear experience, otherwise talk and write as though they are. Readers and listeners naturally assume fear itself is the subject of expressions like "We used freezing as a measure of fear," or "The animals were frozen in fear."

For example, Anderson's work was covered by the press with headlines such as: "FLIES HAVE FEELINGS: FEAR AND MAYBE MORE," and "FLIES EXPERIENCE EMOTIONS LIKE FEAR, AND MIGHT OFFER INSIGHTS INTO HOW THE BRAIN MAKES FEELINGS." This is not what Anderson meant. But regardless of his intentions, the way he set things up pretty much made it inevitable that his findings would be written about this way. I know, since I once did the same thing, and also had to learn the hard way that there is a cost to holding the high ground and thinking, "To hell with the press if they can't get it right."

The idea that threat-elicited behavior in animals is the way to understand conscious fear, and to find ways to treat fear-related psychiatric problems, is not only firmly entrenched in how the press reports it, but also in the way research is conceived and funded. I believe this is undermining the scientific effort to understand fear and develop treatments for fear-related disorders.

In retrospect, I see that the implicit-explicit distinction was doomed to fail since the term *fear* compels the human mind to pattern-complete the concept of fear. While it is possible that more language tweaking could help clarify the semantics of behavioral control, I think a more radical approach is needed, one that allows us to discuss how animals and humans respond behaviorally to significant stimuli in their lives, but without getting confused about, and/or ensnarled in debates about, what they experience when they do so. This conception, I believe, will help pave the way for us to leave ideas about emotional consciousness arising from ancient circuits inherited from animal ancestors behind in favor of a view that builds on the modern science of human consciousness.

*Chapter 63*

# CAN SURVIVAL CIRCUITS SAVE THE DAY?

Every organism has biological mechanisms that underlie activities promoting viability (nutrient and energy management, fluid and ion balance, defense) and fecundity (reproduction). In animals, other than sponges, mechanisms underlying such behaviors are instantiated in the nervous system, specifically in hardwired circuits (figure 63.1). These circuits control Bauplan activities that help keep the organism alive and well, and capable of reproducing. For example, in complex animals a threat to well-being often results in minimization of contact with the potentially harmful stimuli, while the opportunity to obtain needed resources (food, drink, shelter), or to reproduce, typically produces engagement with the relevant stimuli.

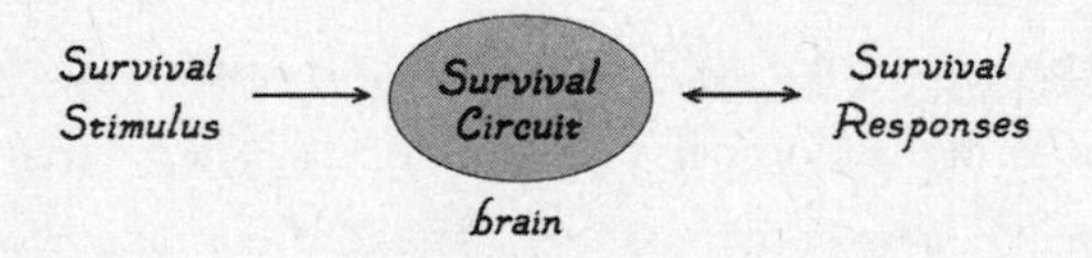


**Figure 63.1:** ***Survival Circuits Control Survival Behaviors in the Presence of Survival Stimuli***

As discussed in previous chapters, the circuits underlying such behaviors are often named with mental state terms—the circuit underlying defense, for example, is typically called a fear circuit, which sometimes leads to the assumption that the mental state named causes the behavior—fear, for example, causes rats and people to freeze or flee when in danger. In an effort to circumvent the confusion that this terminology can engender, I have proposed that the behaviors be called survival behaviors, and the circuits be labeled as survival circuits.

The idea of survival circuits is in a way traditional—it simply refers to the circuits that control hardwired (instinctive) behaviors. Its advantage is that it isolates the objective function (response control) from presumed mental states (feelings). To raise awareness of the survival circuit concept, Dean Mobbs and I recently edited a special issue of the journal *Current Opinion in Behavioral Sciences* (December 2018); we published more than two dozen papers on survival circuits from colleagues in a range of disciplines, including neuroscience, psychology, ethology, and philosophy. In this chapter, I will develop my view of survival circuits, using defensive survival circuits as an example.

When a defensive survival circuit is activated, defensive survival behaviors are triggered, and supported by physiological changes in the body. In the brain, the circuit not only controls motor outputs but also interacts with perceptual, mnemonic, cognitive, motivational, and arousal systems. The body responses feed back to the brain and influence many of the same systems. The net result is the induction of an organism-wide physiological state, a global defensive survival state. This state helps coordinate the various brain and body systems, and supports the organism in its effort to cope with the challenge it is facing (figure 63.2).

The various consequences of survival circuit activity, including the induction of a global survival state, occur to various degrees in all complex organisms with central nervous systems (presumably all or most bilateral animals, including invertebrates and vertebrates). The precise nature of these states will, of course, vary with the complexity of the nervous system, and the resulting behavioral capabilities of a particular animal. For example, in protostome invertebrates, like David Anderson's flies, the global survival state can facilitate innate behaviors (like freezing), as well as learned habits, but in mammals and birds, this state can, in addition, facilitate goal-directed instrumental actions.

Survival circuits, and the global organismic states they engender, control behavior nonconsciously. But in organisms that are capable of conscious awareness of their own brain's activities, the various components of

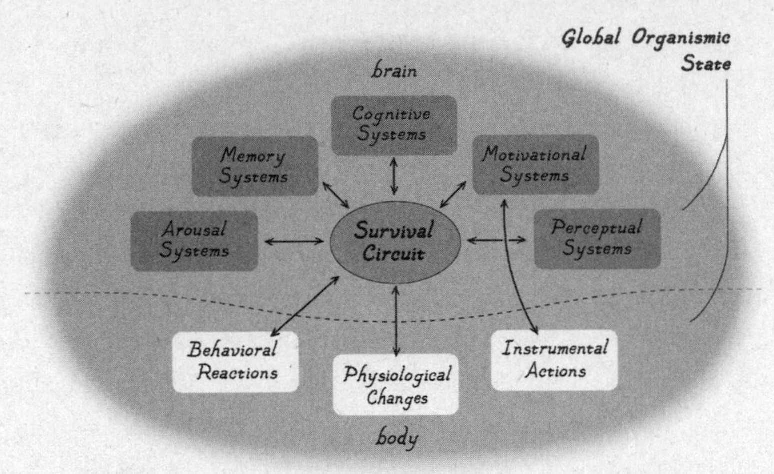


**Figure 63.2:** ***Survival Circuits Interact with Other Systems to Create Global Organismic States***

global survival states can also influence conscious emotions, which, in turn, can result in deliberative control of emotional behavior, including both nonconscious and conscious deliberative control. Emotional feelings can thus have real consequences in our lives. This may seem so obvious as to not need saying, but it does, because some claim that conscious states are epiphenomenal—that they lack actual functional consequences. But to understand their real consequences we have to separate those from the consequences that are inappropriately attributed to feelings.

In the survival circuit model, what I previously referred to as an implicit-fear circuit becomes a defensive survival circuit that initiates an implicit defensive global survival state (compare figure 63.3 with figure 62.1). But, as mentioned in the previous chapter, some scientists want to keep the word *fear* as the name of the state that connects threats to behavior. They insist that they don't mean *fear* in its typical usage, but in the nonsubjective brain-state way—that is, as a scientific construct meaning a physiological state that controls defensive behavior.

Michael Fanselow, who has also contributed much to our understanding of the neural basis of defensive behavior in rodents, views the amygdala as a fear generator, which is a particular version of the nonsubjective

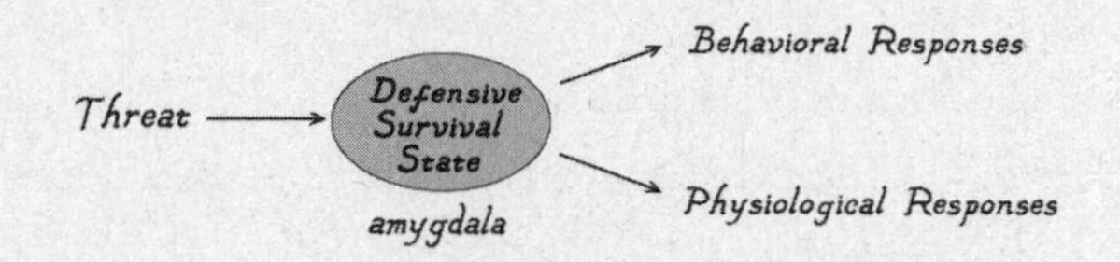


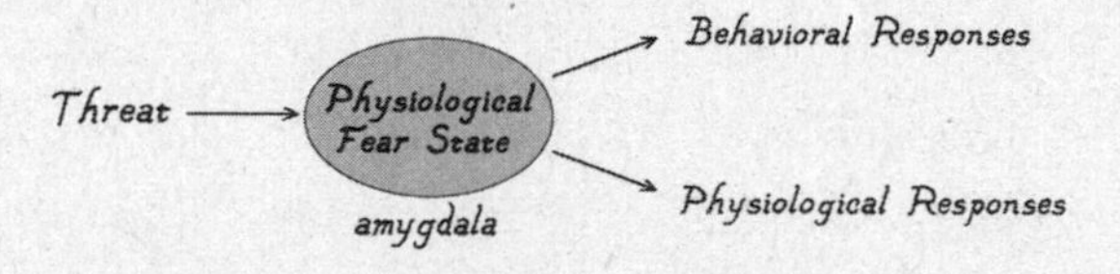


**Figure 63.3:** ***The Survival vs. Fear View of the Amygdala's Role in Controlling Defensive Survival Responses***

fear model of the amygdala. When activated, the physiological fear state results, and controls the expression of defensive behavior. Fanslow allowed for subjective fear, but in his model it is just another consequence of amygdala activity. And compared to other consequences that can be more objectively measured, like behavior, subjective fear is viewed as a crude, imprecise readout of the amygdala fear generator.

The fact is, the defensive survival circuit model and physiological state model, including Fanslow's fear generator model, are quite similar. The main difference centers on what we call the amygdala physiological state that intervenes between the stimulus and the responses it elicits. Should we call it a defensive survival state or a fear state?

Words matter. They underlie our scientific conceptions. If some scientists use words like *fear* to refer to both conscious feelings and to nonconscious states that control behavior, they are obligated to clarify what they mean each and every time they use the word (for example, with modifiers such as explicit and implicit). Otherwise, how are journalists and laypeople, and other scientists who are not privy to the subtleties, supposed to know how to interpret, or even that they need to interpret, the reference? This is especially a problem since these days a number of prominent scientists openly embrace anthropomorphic language and its implications (recall the various efforts described in chapter 41 to make anthropomorphism scientific by naming it with terms like *critical anthropomorphism, biocentral anthropomorphism, animal-centered anthropomorphism,* or *zoomorphism*).

The arbitrary use of words can have significant practical consequences. For example, it has long been assumed that when a medication changes defensive behavior in animals it is because it alters the fear or anxiety circuit, making the animal feel less fearful. A medication that does this should make people feel less fearful, because, it was thought, we have inherited our fear and anxiety circuits from our mammalian ancestors. This effort has been so unsuccessful at finding new medications that some major pharmaceutical companies are ceasing to search for new treatments. Given what we know now, the expectation should have been that medications that target subcortical survival circuits will be most useful in helping people cope with behavioral and physiological symptoms (like excessive avoidance and hyperarousal), which also need to be treated. To change unwanted conscious feelings, though, other treatments that are designed to target cortical cognitive circuits may be required (figure 63.4). With a recalibration of expectations, the field of psychiatry could move forward with an approach to treatment that builds more directly on scientific data rather than presuppositions and assumptions borrowed from nonscientific folk wisdom and commonsense intuitions.

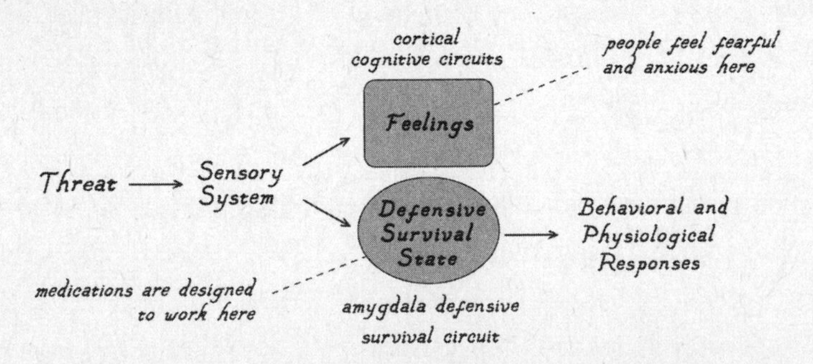


**Figure 63.4:** ***Why Medications Are Not More Effective in Relieving Feelings of Fear and Anxiety***

The best thing for scientists who are not anthropomorphically inclined to do is to be rigorous with their terminology so that they are not assumed to be saying what they are not. Studies of other organisms have an important role in understanding human mental states, including emotions, but only if we have a clear understanding about what findings from the specific organism under consideration can and can't tell us. And some things simply have to be learned from our own kind.

Given that much of the research on fear and the amygdala has used Pavlov's conditioning methodology to study defensive behavior, it is fitting that we let him have the last word here. Shortly before his death, he penned this message to young Russian scientists: "While you are studying, observing, experimenting, do not remain content with the surface of things." These are wise words, and still apt today. We have been too content with surface correlations between behavior and conscious mental states. We should have been digging deeper to understand the conditions under which conscious states do and do not control behavior in humans. And we should have used this information to temper our natural anthropomorphic intuitions about the role of conscious emotional feelings in animals. When all we have to rely on is how animals behave, all we can do is creep up on consciousness.

*Chapter 64*

# THOUGHTFUL FEELINGS

Emotions have been recognized as a key part of human nature since ancient times, perhaps ever since humans could think about their own existence. Plato's notion of emotions as unruly internal "wild beasts" presaged the views of Darwin, MacLean, and basic-emotions theorists, all of whom have assumed that emotions have been inherited from animal ancestors. Aristotle, on the other hand, stressed the importance of thought and reason in emotions, and how these shape moral actions and choices; his ideas influenced a different group of emotion researchers than those aligned with Plato, especially those who came to emphasize the role of cognition in emotion.

I think Aristotle was closer to the truth, but that does not mean that Plato and the Darwinians were completely wrong. We do have instinctual circuits in our brains that control behaviors that occur when we have certain emotions. They just don't *make* those emotions. They are survival circuits, which detect biologically significant stimuli and initiate body responses that help keep the organism alive, continuing the survival imperative that started with LUCA, and that has existed in every biological organism that has ever lived on this planet. Emotional feelings, on the other hand, are, in my view, cognitive interpretations of the situations in which we find ourselves, a capacity that I propose was made possible by the evolution of consciousness.

Cognitive theories of emotion are variants on an idea proposed in 1962 by Stanley Schachter and Jerome Singer. These psychologists argued that an emotional experience is not biologically predetermined, but instead is constructed by the appraisal, interpretation, and labeling of biological, including neural, signals in light of the social and physical context of the particular experience. Although the belief that emotions are hardwired in the limbic system still dominates in neuroscience, the cognitive view of emotions has, thanks to Schachter and Singer, also been a strong force in contemporary psychology.*

For me, human emotions are autonoetic conscious experiences that are cognitively assembled, much like any other autonoetic conscious experience. The idea of unconscious emotion is an oxymoron: If you don't feel it, it's not a feeling, not an emotion. Nevertheless, nonconscious factors contribute.

Schema, you'll recall, are building blocks of cognition. And to the extent that emotions are a type of cognition, schema are crucial to their construction. Pattern completion of schema in the presence of various lower-order, and thus nonconscious, factors drives conscious content. Some of the nonconscious factors include perceptual and mnemonic representations of external stimuli, and representations resulting from activation of survival circuits (figure 64.1).

Schema themselves, as noted earlier, are nonconscious representations that help make sense of the situation in which we find ourselves. And two interrelated kinds of schema are especially important in making conscious emotions—self- and emotion schema.

Being autonoetic experiences, emotional feelings are personal—they crucially involve the self and thus engage one's self-schema. Without the self being part of an experience, the experience is not an emotional experience. Although every experience that involves the self is not necessarily an emotional experience, all emotional experiences involve the self. Lisa Barrett and colleagues note that emotions "cannot be understood independently of an agent conceptualizing him- or herself in a particular situation." The noetic awareness that danger is present is not the same as a state

* Some with cognitive views of emotion over the years have included George Mandler, Richard Lazarus, Nico Frijda, Klaus Scherer, David Sander, Lisa Barrett, James Russell, Kristen Lindquist, Jerome Kagan, James Gross, Kevin Oschner, Andrew Ortony, Gerald Clore, Assaf Kron, Luis Pessoa, Philip Johnson-Laird, Keith Oatley, Rebecca Saxe, among others.

of autonoetic awareness in which you know that *you* are the one in danger.

Also particularly important in the cognitive assembly of conscious emotional experiences are "emotion schema," nonconscious bodies of knowledge about emotions that help us conceptualize situations involving challenges and opportunities. Along these lines, Barrett describes emotions as conceptual acts. Similarly, Gerald Clore and Andrew Ortony note that emotion schema are "ready-made frames" that we use to "interpret the present, remember the past, and anticipate the future." Emotion schema were also central to John Bowlby's attachment theory of child development, and play an important role in Arron Beck's cognitive therapy and variants by Jeffrey Young and Robert Leahy known as schema therapy.

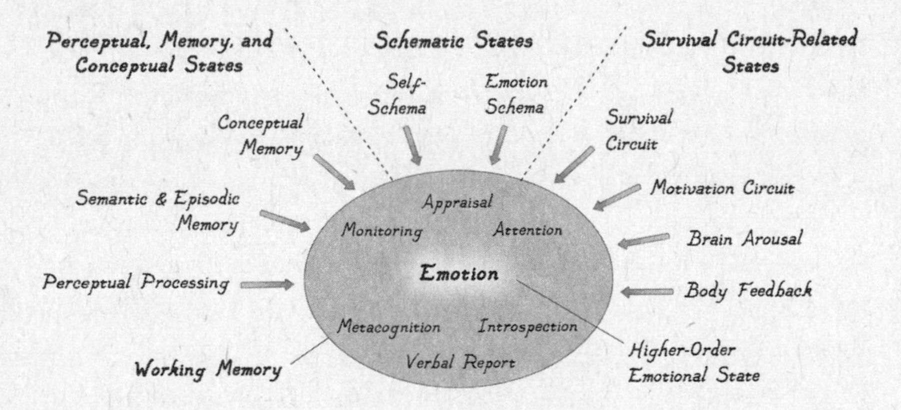


**Figure 64.1:** ***Lower-Order States That Contribute to the Construction of Emotional Experiences***

Jaak Panksepp, who treated primitive basic emotions as products of subcortical limbic circuits inherited from animals, nevertheless proposed that more complex human emotions involve reflective self-awareness and depend on cognitive processing, linguistic representations, and cortical circuits. Antonio Damasio, who also emphasizes subcortical circuits in primitive basic emotions, similarly noted the importance of cognition and language in complex human emotions. But I go further. For me, all emotions, including those typically said to be basic, involve cognitive interpretation based on pattern completion of emotion schema by higher-order circuits.

Your fear schema, for example, is the collection of memories of things that you have learned about threats, harm, danger, and fear itself, including your personal relation to them throughout your life. In the presence of a threat, your fear schema is activated (pattern-completed). The schema then provides a semantic and episodic emotional template that allows top-down conceptualization of the bottom-up, lower-order brain and body states that are also being worked with by cognitive systems memory. The template is a basis for predictive models (expectations) and scripts (possible courses of action) that are typical of such situations. Table 64.1 depicts a verbal representation of a hypothetical fear schema.

TABLE 64.1: Verbal Representation of a Fear Schema

| | | |
|---|---|---|
| aghast | defensive | health |
| agitated | despair | injury |
| agitation | escape | jitteriness |
| alarm | fear | perturbation |
| angst | fearful | phobia qualm |
| consternation | frozen | scare |
| courage | flight | sensitivity |
| cowardice | foreboding | solicitude |
| death | fright | stress |

To the extent that emotion schema are acquired through personal experience, and thereby contribute to higher-order emotional awareness, Axel Cleereman's "radical plasticity theory of consciousness," which emphasizes the importance of past experience to present higher-order conscious states, is worth mentioning here. Also relevant is Barrett's notion of "emotion concepts" that "support categorization and inference, and that control subsequent action." To test this idea, Barrett and Kristen Lindquist placed people in negatively valenced situations that lacked a specific emotional connotation. When the experience of the situation was probed, there was no tendency to label the experience as fear. But if the participants were exposed to stimuli conceptually related to threat and danger first, they were more likely to report that they experienced fear. With Christine Wilson-Mendenhall, Lawrence Barsalou, and others, Barrett proposed that situated conceptualization (conceptual processing relevant in a given situation) shapes the emotional experience that emerges.

Emotion and self-schema overlap considerably. Your fear schema, for example, is yours, having been sculpted by things you have experienced and stored as memories. No one else experiences fear exactly the way you do. What is threatening to you, how threatening it is, and how you are likely to respond are all personal. Some people may pattern-complete a state of fear by simply matching a present stimulus with stored information that identifies the stimulus as a threat. Others may do so only if they determine that escape from the threat is not possible, and the likelihood of harm is high. Still others may feel fear only if they notice an increase in heartbeat, shortness of breath, muscle tension, and/or an abrupt sense of alertness due to an increase in brain arousal. Some may feel nothing in the moment and only find themselves overcome with the feeling of fear as the situation begins to stabilize.

We experience our various emotions differently because each involves a particular schema that contextualizes and interprets the present state differently. Because schema are built up by the accumulation of memories, the earliest emotions one experiences as a child are simpler than the ones that are experienced later. Through assimilation of additional information, a particular emotion schema becomes more complex, and when new information is contradictory, the schema is modified. In this way, a particular emotional state of conscious experience comes to be differentiated from others.

Fear, for example, may start as a rather diffuse, undifferentiated aversive state that, through personal experiences with threats, or observing others in danger, or encountering frightening stories, books, or films, becomes, over the course of time, specific, especially as words are acquired and provide conceptual pigeonholes for distinct experiences. Language may not be necessary for emotional experiences but it certainly helps us differentiate variants of such experiences. Fairy tales and other traditional children's stories, and more recently films and digital games, often depict children in dangerous situations from which they escape unharmed. Such narratives help children build their fear schema and scripts, and at the same time instill confidence that they will be able to deal with threats when they arise.

The popular idea of unconscious emotions needs further elaboration. As I stressed throughout this book, emotions can't be unconscious. On the other hand, because nonconscious schema are building blocks of conscious emotional experiences, feelings can seem to reflect nonconscious emotions. And since schema also influence behavior, actions can seem to have been driven by a nonconscious emotion. But emotion schema are not emotions—they are the cognitive launchpads of emotions.

If the actual experience of a given emotion varies from person to person, why do certain emotions, fear being a good example, seem so universal? What is actually universal about fear is not the details of how it is subjectively experienced, but rather the concept of fear. All organisms face threats to physiological and/or psychological viability. These are among the most significant stimuli they will encounter in life. As a result, organisms that can be conscious of their brain states will typically have a conscious experience when threatened. And if the organism has language, this experience will certainly have been given a name, as information about what is dangerous is certainly important to share with others in one's group. Consequently, every speaking culture will have the concept and a word, or even many such words, for states that occur when its members face danger. Ultimately, what these concepts and words actually mean to an individual will be shaped by her personal experiences.

Emotions and their labels come in families, with a central core (fear) and peripheral exemplars (concern, anxiety, dread, terror, panic). As an emotion schema develops over time, the emotion terms used to describe relevant experiences become more refined—"I was terrified" expresses a more intense feeling than "I was afraid" or "I was scared." Adverbial support contributes further refinement: adding "really" to "scared" moves it toward "terrified." As experiences accumulate, emotion words become emotion concepts, and emotion concepts become schema that can be used to categorize new experiences.

The proposition that language contributes to emotion is often challenged. For example, a common criticism is that you only have to watch preverbal babies to see their emotions on display, despite the fact that they cannot speak. But responses like these are *not* unambiguous indicators of feelings, in part because, as we have seen, responses and feelings are controlled by different circuits in adults, and the circuits that control the responses mature earlier in life than the cognitive circuits that underlie conscious experiences. As in other animals, the absence of speech is a major hurdle in demonstrating consciousness in preverbal children based on behavior alone.

Infants gradually discover regularities in their perceptual world and begin to form nonverbal memories, including schema, about its workings. By the time a child can speak, she thus has at least the beginnings of a conceptual structure on which to base speech. That said, preverbal cognition may well support some form of noetic awareness about danger before the brain has matured to the point of having language and self-awareness. But that is not fear. No self, no fear.

So far, we have considered emotion language in terms of semantic labels, such as fear, joy, sadness, anger, and the like, an approach I also took in *The Emotional Brain.* I was later criticized for this by Zoltán Kövecses in *Metaphor and Emotion.* He argued persuasively that emotion language should not be viewed as a collection of literal words that categorize and refer to a preexisting emotional reality, but rather as figurative and metaphorical concepts that define and even create emotional experiences for us. Metaphors, in other words, become components of schema and are used to conceptualize and organize our inner and outer worlds.

Emotions are, in the moment of experience, incorrigible (unmistakable). If I have an experience that is pattern-completed as fear, fear is what I feel, regardless of how I outwardly appear and/or act, and regardless of what others think I am feeling. If I later revise my view, and decide instead that I was angry or jealous, I am rewriting my personal psychological history from memory. But not all significant experiences are pattern-completed in a way that necessarily has a precise native-language, standard-emotion term associated with it—for example, diffuse, vague, amorphous states of aversion might be labeled with general terms such as unease, discomfort, or distress, and in more positive situations as comfort, contentment, elation, or well-being. These too might be revised or updated later by applying a conventional emotion word—unease may become anxiety, and elation happiness.

As Gerald Clore, Daniel Kahneman, and others have noted, the closest we ever come to the truth of experience is when we are in the experience. Everything that happens later is a top-down reinterpretation of the memory of the original experience. In therapeutic situations, such revisionist histories can be revealing of one's psychological tendencies and proclivities, but they can never be as true to a past experience as the experience itself was in real time. The mere act of retrieving a memory can, through the process of memory reconsolidation, change the nature of the memory (this is a natural process through which memories are changed after retrieval, requiring restorage). The longer the amount of time that passes between the actual experience and the memory of it, the greater opportunity to revise and reconsolidate it upon retrieval of the remembered narrative.

Self-narrative revision is indeed common. Clore and Andrew Ortony argue that this is an important function of emotion schema, noting that when people "recall and retell their experiences, they necessarily edit, embellish, and assimilate them to whatever categories of understanding are available. Such retellings presumably benefit from an implicit library of emotional schema that help both speakers and listeners make sense of events." Interestingly, recent research suggests that an effective way to help people with trauma is to have them write about it. So-called writing exposure therapy (WET) is a variant of cognitive behavioral therapy. Writing about trauma is proposed to modify and clarify one's trauma narrative, and provide some relief from the enduring effects of the trauma. Research suggests that WET may achieve results faster than both traditional cognitive behavioral therapy and medications. We understand ourselves through the stories we tell, both to ourselves and others, about who we are.

*Chapter 65*

# EMOTIONAL BRAINS RUN HOT

Despite the importance of emotions to human mental life, theories of consciousness have been relatively silent on this topic. But if emotions are, as I propose, products of the same general cortical cognitive circuits that generate other kinds of conscious experiences, we can leverage what we know about consciousness from studies of perception and memory to begin to understand emotional experience. With this idea in mind, Richard Brown and I outlined a higher-order theory of emotional consciousness, using fear as an example. Here I extend our model in light of the multistate hierarchical model I have been developing.

The basic idea is that the conscious experience of emotion results from the higher-order representation of nonconscious lower-order states by the higher-order network (dorsal and ventral lateral prefrontal cortex and frontal pole). This network, I propose, is key to the experience of not just perceptions, memories, and thoughts, but also of emotions. However, there is clearly something different about states we label as emotional and those we do not. That difference, I believe, is defined by the kind of information used by the higher-order network when it makes emotions, as opposed to other kinds of experiences.

Figure 65.1 shows three classes of states I think contribute to conscious emotional experiences in the presence of emotional trigger stimuli; for example, a threat to well-being in the case of fear. The first two occur in both emotional and nonemotional conscious experiences. These include states related to noetic awareness of trigger stimuli (perceptual, memory, and conceptual states), and states related to autonoetic awareness of one's

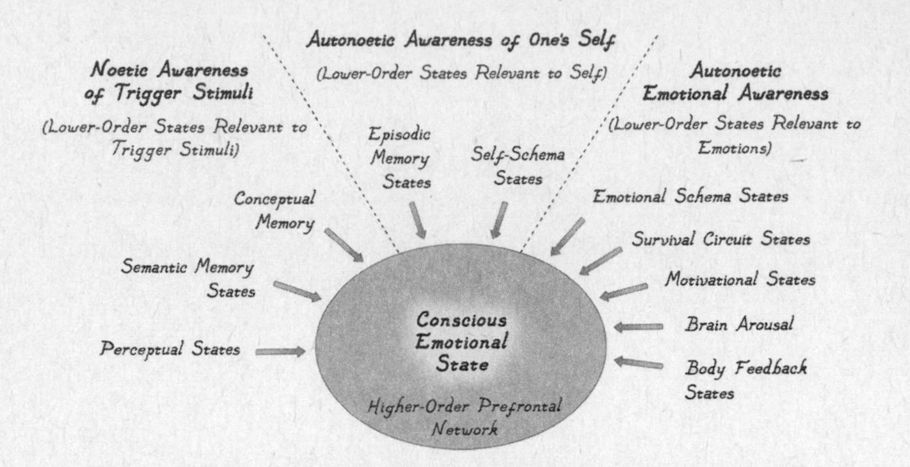


**Figure 65.1:** ***Lower-Order States that Contribute to the Construction of Noetic, Autonoetic, and Emotional Autonoetic Awareness in the Presence of Emotional Trigger Stimuli***

self (episodic memory and self-schema states). The third class involves states that help define the quality of the emotional experience itself and include emotion schema, survival circuit activity, motivational activity, arousal, and body conditions.

The goal of this chapter is to explore how these various states are used in the neural assembly of autonoetic emotional experiences by the higher-order prefrontal network. But to do this, we have to also include a second prefrontal network involving medial areas (orbital, anterior cingulate, ventromedial, and dorsomedial) and the anterior insula. These additional prefrontal areas receive many of the same inputs as, and connect with, the higher-order network. They are thus sources of complex cognitive representations that can, like perceptual, mnemonic, and conceptual representations, be used in the assembly of conscious experiences by the higher-order network (figure 65.2). As such, despite the fact that they are in the prefrontal cortex, they are lower-order with respect to the higher-order network of consciousness—they are, in other words, components of the hierarchy of lower-order states re-represented by the higher-order network.

I will illustrate the way these various lower-order states are used in the construction of an emotional experience by tracing the fate of a threat

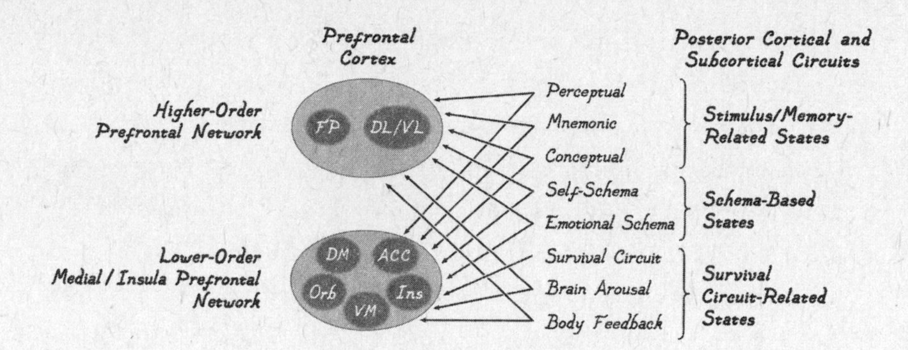


**Figure 65.2:** ***Lower-Order Input Signals to the Higher-Order and Lower-Order Prefrontal Networks***

stimulus as it engenders an autonoetic state of conscious fear. The process starts when information about the trigger stimulus—say, a snake at your feet—is transmitted from your eyes to the primary visual cortex. Secondary visual circuits, some using memory of past perceptions to filter signals, then distribute their outputs to prefrontal areas (especially the dorsal and ventral lateral prefrontal cortex). Secondary sensory circuits areas also send outputs to circuits that add additional semantic and conceptual meaning to the representations (including the medial temporal lobe and neocortical areas, such as the temporal pole, among other multimodal regions). These latter circuits send their outputs to the medial/insula prefrontal areas, which connect with the higher-order network; some also connect directly with the higher-order network. On the basis of the various prefrontal representations by the higher-order network, top-down control over processing is initiated and influences ongoing posterior perceptual, mnemonic, and conceptual processing. As these processes unfold, interactions within the higher-order network begin to shape a perceptual conscious experience of the threatening stimulus and its context.

At this point, you have thus achieved a noetic state of consciousness—an awareness that harm is present—but are not in an autonoetically conscious emotional state—one in which you are aware that *you* are in the presence of harm. But what good is noetic awareness of a threat as a conscious state if it is not a state of emotional consciousness? As discussed in part 9, states of noetic consciousness offer a high level of deliberative cognitive control over instrumental behavior. An animal that can consciously control such behavior on the basis of stored representations of value has processing advantages, and especially decision-making advantages, over one that is limited to nonconscious cognitive deliberation, and especially over one limited to trial-and-error instrumental learning.

Given that a noetic experience of a threat lacks a personal element, and is thus not an emotional experience, how does the brain go further and construct an autonoetic emotional state? Self-referential information is, of course, necessary for a higher-order experience of one's self. But not all such experiences are necessarily emotional ones. Self-representations can be purely cognitive (your personal knowledge that there is a pencil present and that it is *you* who is looking at this pencil is not an emotional experience, unless you have been harmed by a pencil). By contrast, other self-representational experiences *are* emotional (*you* fear that harm will come to you when you see a snake at your feet). Additional lower-order ingredients are necessary to transform a purely cognitive autonoetic state into an autonoetic emotional experience. Particularly important are survival-circuit-related activities.

By way of sensory inputs from thalamic sensory areas and secondary cortical sensory areas (the so-called low and high roads to the amygdala, respectively), threats activate the defensive survival circuitry, and initiate a cascade of events in the brain and body. The amygdala has direct and/or indirect connections with many cortical areas and influences information processing in them. Direct connections with the sensory cortex biases and facilitates sensory processing there. Connections with memory areas in the neocortex and medial temporal lobe facilitate the retrieval of semantic memories of specific items, concepts, and schema that support object recognition and situational classification, and also facilitate retrieval of episodic representation of past experiences. Connections from the amygdala to the medial and insula prefrontal areas allow the amygdala to indirectly affect interactions with the lateral prefrontal executive and higher-order circuits, and influence top-down conceptual control over sensory processing, memory retrieval, and decision making, as well as the higher-order construction of emotions.

Amygdala survival circuit activity also leads to the activation of neuromodulatory systems that raise levels of arousal throughout the brain. For example, amygdala-initiated arousal facilitates processing in sensory, memory, and prefrontal circuitry, leading to vigilant scanning of the environment for clues about the cause of the arousal, with retrieved semantic memories focusing the search. Amygdala-initiated arousal also affects the amygdala itself, creating a wave of feed-forward activity that facilitates all other amygdala-initiated activities, so long as the threat is present.

The most widely discussed effect of amygdala survival circuit activity is the triggering of behavioral and physiological body responses. These innate reactions serve as the organism's initial, immediate way of coping with the potential source of harm detected by the survival circuit. The body responses feed back to and affect processing in the brain. For example, survival-circuit-triggered body responses produce sensations that are processed in what Damasio calls body-sensing areas of the brain. One such area is the insula cortex, which also receives signals from the amygdala, and like the amygdala, connects with the medial prefrontal cortex, and the insula and medial prefrontal areas connect with the higher-order prefrontal network. In addition, release of adrenaline from the adrenal medulla in response to survival circuit activity leads to activation of nerves in the body cavity that send signals to neuromodulatory systems in the brain, further enhancing brain arousal triggered directly by the amygdala. Cortisol released from the adrenal cortex travels in the bloodstream to the brain, where it affects processing in many of the cortical and subcortical areas we have been discussing. Although hormonal effects are slow to take effect, they also inactivate slowly.

The net result of all this amygdala survival-circuit-initiated activity is what I earlier called a global defensive survival state. By way of loops of connections between the various players, the global defensive state becomes self-sustaining: the defensive reaction (e.g., freezing) either continues or gives way to the motivation of instrumental actions (e.g., escape or avoidance) and enables deliberative responses that persist as long as the threat remains. The slow onset and offset of hormonal secretions means that a defensive survival state and its outward manifestations can continue even after the threat itself has dissipated—one may feel "shaken" or "jittery" for some time after encountering a snake or being mugged. Hormones can have even more persistent effects as a result of their strengthening of both explicit and implicit memories that are formed during stressful encounters and carried forward in time.

Given the extensive direct and indirect influence of the amygdala on cortical processing, it would be tempting to conclude that its activation is the defining condition for the conscious experience of fear, and could therefore be used as a biomarker of fear. Although this idea is popular, and is often assumed to be the case by scientists and laypeople, it is, as noted earlier, problematic for several reasons that need to be reiterated. First, the mere activation of the amygdala does not necessarily mean that the defensive survival circuit has been activated as well—the threat-sensitive defensive survival circuit is one of the many circuits that pass through the amygdala; some others are those involved in appetitive states related to food, drink, or sexual partners. Second, even when the amygdala defensive survival circuit is activated by a threat, fear does not inevitably occur—recall that subliminal threats activate the amygdala defensive circuit and elicit defensive responses, but do not elicit fear. Moreover, even when the experimenter asks the participants to guess what feeling they might be experiencing, they come up short. Third, amygdala activity is *not* required to elicit fear, since fear can be experienced when the amygdala is damaged. Fourth, the amygdala defensive survival circuit is not the only survival circuit that can support fear—one can fear death from starvation, dehydration, or hypothermia, each of which is managed by other survival circuits. Such lower-order survival circuit states thus contribute to fear, but are not the defining factor in fearful experiences.

If not the amygdala, then what? The answer, as you probably know by now, involves your fear schema, activation of which is crucial for defining the fearful nature of an experience. Fear is the result of any condition or set of conditions that activates elements of your unique fear schema, giving rise to a cognitive state in which you become consciously aware that you are in physical or psychological jeopardy. What is threatening to you is not necessarily of concern to others, or perhaps not to the same degree of concern—our experiences of environmental, body, or brain states are always, to some extent, personal.

To summarize, my proposal is that a conscious emotional experience typically results from the processing of various nonconscious, lower-order ingredients by the prefrontal higher-order network: (1) perceptual information about the triggering event; (2) retrieved semantic and episodic memories; (3) conceptual memories that add additional layers of meaning; (4) self-information via self-schema activation; (5) survival circuit information; (6) brain arousal and body feedback consequences of survival circuit activation; and (7) information about what kind of emotional situation might be unfolding as a result of activation of one's personal emotion schema.

The higher-order network attends to, monitors, and controls processing of these nonconscious lower-order signals and uses them to introspectively access, label, and experience the resulting autonoetic conscious emotional state. If your fear schema has been pattern-completed by a threat, the experience will fall in the general domain of fear, and a word from the fear family lexicon that is available to you will very likely be used to label it as fear, panic, terror, anxiety, worry, or concern. The schema elements activated define the experience in the moment; the label simply refines and anchors it.

We sometimes hear about the possibility of experiencing two or more emotions at the same time. But I think a better way to think about this is that emotions can be labile. If due to the complexity of a situation, multiple emotion schema, or rapidly changing schema elements, are being activated around the same time, different ones may rise into awareness at different moments. In other words, as circumstances change from moment to moment, inputs to the higher-order network will also change, and the momentary experience can also change—concern can give way to panic, terror, or some other variant of fear, and these can morph into anger or love, affection or rage, joy or sadness—all depending on which schema elements momentarily dominate. While you can't experience fear and love or anger simultaneously, if different schema elements are rising and falling in activation during some episode, a psychological illusion of simultaneity may result.

A key prediction of the theory is that damage to the amygdala should either have no effect, or should only dampen rather than eliminate feelings of fear. Pattern completion of a fear schema should, on its own, be sufficient to produce memory-based representations that mimic the flavor of a fearful emotional experience, if not the intensity of more typical fearful experiences that include body feedback. However, it is also important to note that once you have conceptualized that you are afraid, this awareness may, in top-down fashion, be a sufficient trigger for initiating the kind of brain and body arousal typically elicited by external threats. This idea is also captured in Damasio's proposal of an "as-if" loop—a kind of mental simulation or predictive inference that can substitute for missing lower-order signals.

Recent studies by Lisa Barrett and others have begun to demonstrate the importance of top-down control, predictive coding, and active inference in emotional processing and experience. And, as I have noted previously, the idea that top-down predictions and inferences influence conscious emotions can be viewed as compatible with a higher-order view. For example, in a higher-order account, especially a HOROR account, missing body feedback representations can be thought of as absent lower-order states that are made up for by a top-down nonconscious conceptualization in the form of a mental model/schema.

It also important to recognize that some emotions don't require the activity of survival circuits and the consequences they produce; for example, so-called secondary, or social, emotions (e.g., guilt, jealousy, shame, embarrassment, pride, contempt) and existential emotions (e.g., dread over the meaningless of one's life). But in my model, all emotions (whether basic, secondary, or existential) are cognitively assembled states of autonoetic consciousness. As such, they are all products of the same higher-order circuits that underlie all varieties of autonoetic conscious experiences, not just emotional ones. When survival circuits are part of the mix, they modulate the experience but do not determine the experience, except to the extent that they help pattern-complete emotion schema elements.

The contribution of the higher-order network, and its processing of inputs from other prefrontal and posterior multimodal areas to conscious emotional feelings, has not been explicitly studied in detail. Existing data have implicated many of the regions in aspects of emotional processing, and, in some cases, in emotional experience. But it clearly remains for future research to specifically test their contribution in the context of a higher-order neural account of emotion.

It is also possible that the higher-order network is not rigidly fixed. The various prefrontal areas, or even posterior areas, might be recruited on a situational basis as a momentary higher-order coalition (not unlike the notion of a working memory coalition described earlier). This could be especially important when some specialized processing might be needed, or in situations in which areas of the higher-order network have suffered injury.

I am obviously a proponent of the higher-order theory of consciousness, but I also recognize that exactly how consciousness arises in the brain is still an open question. Whatever the answer (a variant of HOT, global workspace, or some other theory), it will, I believe, be a general theory that accounts for emotional and nonemotional conscious states alike within one framework.

*Chapter 66*

# SURVIVAL IS DEEP, BUT OUR EMOTIONS ARE SHALLOW

No matter how simple or complex, all organisms sustain life and contribute to species persistence by managing energy resources, balancing fluids and ions, defending against harm, and reproducing. These fundamental survival activities are instantiated in dedicated circuits that control specific innate behaviors in organisms with central nervous systems. But these circuits do not make emotions.

I propose that emotions are human specializations made possible by unique capacities of our brains. They could not exist in the form we experience them without our early hominid ancestors having evolved language, hierarchical relational reasoning, noetic consciousness, and reflective autonoetic consciousness. These capacities made it possible for activities of ancient survival circuits to be integrated into self-awareness, framed in terms of semantic, conceptual, and episodic memories, interpreted in terms of personalized self and emotion schema, and used to guide behavior in the present and also to plan for future emotional experiences. Emotions thereby became the mental center of gravity of the human brain, fodder for narratives and folktales, and the basis of culture, religion, art, literature and relations with others and our world—of all that matters in life as we know it.

But emotions, rather than being an inherited vestige of our primate or mammalian past, may be exaptations that reflect unique features that first emerged in early members of our species. Exaptations, you'll recall, are useful traits that arise as by-products of other traits, and, because of their value, come under genetic control through natural selection.

Emotions might, in fact, be the result of two other exaptations. One was language, which, as noted earlier, Kolodny and Edelman suggested arose from synaptic plasticity that coupled neural mechanisms underlying nonverbal communication, serial cognition, and tool use. But in addition, by providing personal pronouns, language may well have enabled another exaptation—autonoesis (awareness of "self as subject," as opposed to "self as object"). Emotions, being a form of autonoesis, were then inevitable as conscious experiences of one's self in *biologically or psychologically* significant situations in life. The biologically significant ones are most typically associated with survival circuit activity, while the psychologically significant ones are those that lack a fundamental connection to survival circuits. What makes them all part of the category we call emotions is thus not that they possess some biological signature, but instead the fact that they are of personal significance to one's self.

That human emotions might be exaptations that were later selected does not mean that they have no connection to our animal ancestry. Indeed, the most fundamental emotions are those to which ancient survival circuits contribute. But, as I have argued, these survival circuits influence, but do not define, the content of emotional experiences.

The utility that allowed emotions to persist in the genes of our species may have been the ability to personalize value. Rather than simply detecting risk and avoiding danger, the organism could consider, "How dangerous is this to *me*?" Other animals can represent value, but only humans can make it personal. In this view, an emotion is the experience that something of value is happening to you. If so, emotions could not exist without autonoesis. No self, no emotion.

Since Darwin, scientists have struggled to connect human behavior to the history of life by assuming that emotions are a key link. The flaw in the Darwinian emotion doctrine is revealed once we recognize that the behavioral survival capacities we have inherited from our animal ancestors are products of different brain systems than those unique ones that make emotions and other states of autonoetic consciousness in humans. The historical waters of survival behaviors are deep, but the stream of emotional consciousness is shallow.

The idea that important aspects of human brain function are new does not reduce the status of other animals to that of primitive reflex machines. Even humans, the only organisms for which we have clear evidence of self-awareness, go through much their of day-to-day business using sophisticated nonconscious cognitive and behavioral capacities, many inherited from our animal ancestors. Existing research suggests that animals and humans are indeed quite similar, not because animals have human consciousness, but because humans have inherited nonconscious capacities from them. Gaining an understanding of the animal heritage of nonconscious functions of the human brain is thus not a consolation prize. It is crucial for our understanding of both animal and human behavior.

The romantic idea that other mammals are furry, primitive people with the full complement of human psychological features (despite not having the full complement of human neural features) is intuitively compelling, and works well in our day-to-day interactions with our pets, but I believe misguided as a scientific notion. Once we accept this, the task of connecting humans to the deep history of life is greatly simplified. It allows us to understand the relatively primitive capacities that guide behavioral commerce with predators and other dangers, and with foods, fluids, and sexual partners, in similar terms in humans and other animals without having to wade in the muddy scientific waters of animal consciousness. These behaviors are accounted for by conserved survival circuits. They do not require that we postulate and search for mental states that scientifically we cannot readily assess in other animals. Perhaps some animals have the capacity for noetic consciousness. But if, as I contend, emotions are autonoetic states, they may be ours alone.

Our survival circuits connect us to the survival history of organisms with nervous systems. And the universal survival strategies that survival circuits and behaviors tactically implement connect us to the entire history of life. Separation of the history of emotions and other states of consciousness from the deep history of survival circuits allows us to see our place in this ancient story.

Like all other species, we are special because we are different. Our differences are important to us because they are ours. But they are mere footnotes in a four-billion-year-old saga. Only by knowing the whole story can we truly understand who we are, and how we came to be that way.

*Epilogue*

# CAN WE SURVIVE OUR SELF-CONSCIOUS SELVES?

Like all living things, humans are organisms, biological entities that function as physiological aggregates whose constituent parts operate with a high degree of cooperation and a low degree of conflict. But unlike other organisms, humans possess a rogue component—a brain network that can, at will, choose to defect and undermine the survival mission and purpose of the rest of the body. This is the network that underlies human consciousness, and especially our capacity for autonoetic, reflective self-awareness.

Autonoetic consciousness (the ability to mentally model one's self in relation to time) is the essence of who each of us is, or at least of what we consciously know about ourselves. It is the basis of the conceptions that underlie our greatest achievements as a species—art, music, architecture, literature, science—and our ability to appreciate them. For good reason, then, consciousness researcher Hakwan Lau calls his blog *In Consciousness We Trust.*

But should we? Consciousness, especially autonoetic consciousness, has a dark side—it is the enabler of distrust, hate, avarice, greed, and selfishness, mental features that could be our undoing.

Wait a minute. Isn't survival, life itself, an exercise in selfishness? Isn't selfishness the way organismic unity is maintained? Isn't it at the heart of Richard Dawkins's "selfish gene" theory of survival? Aren't bacteria and bees and worms, fish, snakes, cats, and apes selfish? The answer to all of the above is yes—how else could they survive? But something unique happened when selfishness came to be an isolated capacity in humans—that is, when selfishness became the basis for conscious decisions that could harm, rather than simply enable or enchance, the well-being of the organism as a whole.

Unicellular organisms had the Earth to themselves for more than three billion years. Multicellular organisms evolved by transferring responsibility for fitness and survival from the single cell to a more complex entity with many cells that all shared a common genome. This biological model worked fairly well for nearly another billion years, until organismic unity was suddenly challenged by the arrival of the capacity for autonoetically conscious brains in humans.

The autonoetically conscious human brain is the only entity in the history of life that has ever been able to choose, at will, to terminate its own existence, or even put the organism's physical existence at risk for the thrill of simply doing so—the other cells and systems be damned. Some argue, on the basis of anecdotal evidence, that other animals also commit suicide. But whether such behaviors are truly intentional, in the sense of being based on a thought about causing one's self to cease to exist, is controversial. The famed late-nineteenth-century sociologist Emile Durkheim proposed that suicide applies only to cases of death resulting directly or indirectly from a positive or negative act of the victim himself, an act which he knows or believes will produce the intended result—death. Because this kind of conception depends on a reflective social consciousness, animals, with their purely internal, physiological constraints, are incapable of it. True suicide, in its various forms, according to Durkheim, is a social condition of humans.*

Early humans are believed to have been unremarkable compared to coexisting fauna. Then, at some point (estimates range between fifty thousand and two hundred thousand years ago), something happened to distinguish our ancestors from the rest of the animal kingdom. They developed new capacities and ways of existing and interacting with one another—language; hierarchical relational reasoning; represen-

* Thanks to Hakwan Lau for alerting me to Durkheim's views. My summary of Durkheim's position is based on Robert Alun Jones, *Emile Durkheim: An Introduction to Four Major Works.*

tation of self versus other; mental time travel. Autonoesis was the result.

That autonoetic consciousness might be unique to humans does not mean that it appeared out of the blue. For one thing, our primate ancestors had sophisticated working memory capacities, including executive functions, which allowed the integration of perceptual and mnemonic information and nonconscious deliberation about alternative courses of action. These were made possible by their lateral prefrontal areas (including dorsal and ventral lateral areas), which present-day monkeys and apes also possess, but their nonprimate mammalian ancestors lacked. More speculative is the possibility that these circuits may have also made it possible for ancestral primates to have noetic conscious experiences of perceptual events, and perhaps could experience a kind of noetic awareness of the value of such events based on a crude semantic appraisal of gradations of distinctions between what is generally useful and harmful. Perhaps they could even experience a relatively simple noetic version of self-awareness based on semantic autobiographical information about what belonged to their body and what did not.

But they would *not* have been able to experience their selves as entities with a personal past, and imagine their selves in possible future manifestations, including the existential realization of eventual nonexistence. Autonoetic awareness, I propose, depended on unique features that we know typify the human prefrontal cortex: a region, the frontal pole, that has novel components and that interacts with lateral prefrontal areas to form the higher-order network; enriched connections between the higher-order prefrontal network and lower-order processors (including other prefrontal areas and perceptual, mnemonic, and conceptual processors in the occipital, temporal, and parietal lobes); and novel cell types and molecular/genetic mechanisms that fostered enhanced processing within the higher-order network and between it and lower-order processors.

Given that autonoesis can pose a threat to organismic unity by undermining the overall survival goals and needs of the organism, it must have added significant survival value that protected it from being eliminated as a passing evolutionary fad. One obvious possibility is that with autonoetic consciousness, rather than simply detecting risk and avoiding it, the organism can personalize the risk by, for example, asking, "How dangerous is it to *me*?" This, I believe, is how autonoesis made emotions possible. An emotion is the experience that something of value is happening to you. No self, no fear—nor other emotions.

While many animals make decisions based on value about what is useful and harmful, only humans dynamically evaluate the implications of a situation in real time using complex hierarchical decision trees to draw conclusions and take actions relevant to personal well-being in the moment. Only humans plan for an imagined future, or even a set of alternative possible futures.

And only the autonoetically conscious human mind can generate a narrative in which the danger of a certain option, after having been calculated, can be counterfactually minimized so that a selfish desire can be satisfied, free of guilt or anxiety during the act—consuming a rich, high-caloric meal, swimming in rough seas, scaling the walls of a cliff, having an adulterous fling, or taking an addictive drug.

Autonoesis is a double-edged sword. Our future depends on how we, as a species, choose to use it.

Earlier I quoted Aldous Huxley as saying we rose above the brutes because of language. He also maintained that people can easily become victims of their words. Language gives us personal pronouns that separate "me" from "you" and "us" from "them." We build social groups, clans, tribes, religions, kingdoms, and nations on this basis, and shun, isolate, harm, and even kill one another to protect beliefs that define the groups with which we choose to affiliate. The selfishness of our genes pales next to the selfishness of our self-conscious mind and its convictions.

Beliefs are not just products of language or culture. They also depend on other special capacities that are intricately entwined with language—hierarchical cognition, self-awareness, and emotions. When these blend seamlessly, social systems that work for the greater good of our kind are possible. But when emotions are at odds with our reasoned thoughts, or when either is corrupted by beliefs, or when personal interests are pitted against the values of the culture at large, or against the needs of our species as a whole, humans suffer.

The personal, selfish nature of the autonoetic mind leads it to assume that it is always in charge. Indeed, so-called free will is one of our most cherished narratives, which, according to the Bible, began when Adam chose the apple. And since ancient Greece, humans have believed that we are our conscious minds; we have treated the rest of the mind/brain and the body as servants, mere support staff. Descartes's dualistic philosophy was an attempt to reconcile religious conceptions of the soul in light of the scientific revolution begun by Copernicus and Galileo. The philosopher Søren Kierkegaard later proposed that anxiety is the price humans pay for the freedom to consciously choose how to lead our lives. While the behaviorist movements attempted to eliminate consciousness as a scientific construct, consciousness itself did not let that rejection stand.

Our unique brains have enabled us to conquer frontiers. We have the power to change the environment to meet our needs; satisfy our whims, desires, and fantasies; and protect ourselves from our fears and anxieties. Imagining the unknown inspires us to find new ways of existing.

Our thirst for knowledge has led to scientific and technological discoveries that have made life, at least for the lucky among us, easier in many ways. We don't have to forage for food in dangerous settings—bloodthirsty predators are simply not part of daily life for most humans. We easily combat seasonal changes in temperature with convenient appliances. We have access to medications to treat, and even prevent, common illnesses, and surgical procedures can fix and, in some cases, replace damaged body parts.

And we can electronically communicate with people anywhere in the world instantaneously. The internet has transformed life in ways worth celebrating, but like most good things, it comes at a cost. It has made it easier to be self-centered by facilitating realignments of interests that oppose the common good, challenging commonly accepted beliefs through hearsay and rumor, and even outright lies. False assertions can gain credence simply through repetition. Such tactics have been used to undermine the value of science and its contributions to life and well-being, and to attack the foundations of our social structures, including our government, and its safety nets for those in need, and its checks and balances against tyranny.

In the past the pace of change was slow and incremental, but over the last century it has become fast and furious. Global temperature is rising, along with unusual weather patterns. Forests are burning. Deserts are expanding. The seas are rising. The rate of species extinction is accelerating.

Many alarmed observers have called for efforts to "save the planet" by reversing, or at least slowing, the changes we have wrought. Others, though, have been swayed by the belief system of climate change deniers who insist that the relevant research is a hoax.

The astrophysicist Adam Frank believes that those concerned with our current situation are right to worry. Human actions, he says, are indeed having adverse consequences, and, he argues, are on target to drastically modify the physical and biological constitution of the Earth. But we won't destroy it. Quoting Lynn Margulis, originator of the endosymbiotic theory of multicellular life, Frank says: "Gaia is a tough bitch." Our planet, Frank reminds us, has survived significant geophysical disasters and mass extinctions in the past and will persist. But if we don't make corrections soon, it may not persist in a way that will support the current configuration of organisms, including us.

Bacteria and archaea, the ultimate survivors, will surely make it. Large multicellular organisms with voracious energy appetites may have a harder time. We know from past mass extinctions that opportunities arise for those who survive. The biological experiments that result will likely create a very different profile of life on Earth. And without us mucking around the way we do, the natural order of things might reach a more stable equilibrium. The philosopher Todd May, pondering such issues, recently asked, "Would human extinction be a tragedy?" He concluded that the world might well be better off without us. But his key question was, Would a world without our kind be a tragedy, given that we have achieved such remarkable things as a species?

Autonoetic consciousness is ultimately personal and selfish, and at its worst moments, narcissistic. Self-consciousness, according to Christophe Menant, is also the root of evil. At the same time it may be our sole hope for a future.

With our autonoetically conscious minds, we have constructed conceptual guidelines, such as morality and ethics, to help make difficult decisions, for example, about our way of life. Only self-conscious minds can come to the realization, as Todd May's mind did, that we have an obligation to confront our selfish nature for the good of humankind as a whole. But in the end, this is a value judgment, one based on the assumption that our achievements are special.

Autonoesis allows us to care about our differences, and bemoan their possible demise. There's nothing wrong with that. But perhaps we can sustain some version of our way of life without asking too much from other organisms. Doing so might well avert drastic changes in the configuration of life—the balance of biological power—that climactic change can bring. Remember, small mammals with low energy needs rose to the top of the food chain when conditions became less favorable for larger, energy-demanding, reptilian predators that had dominated with abandon.

We persist as individuals only if we persist as a species. We don't have time for biological evolution to come to the rescue—it's too slow a process. We have to depend on the more rapid avenues of change—cognitive and cultural evolution, which, in turn, depend on our autonoetic brains and their choices. In the end, it is indeed consciousness in which we must place our trust.

# APPENDIX

## Timelime of the History of Life

(all dates approximate)*

4.6 ***bya*** Earth formed

4.0 Early life experiments

3.8 LUCA (ancestor of all present-day life)

3.5 Prokaryotes (bacteria followed by archaea)

2.0 Eukaryotes (unicellular protists)

1.2 Common protist ancestor of plants, fungi, and animals

1.0 Specific protist ancestor of plants, fungi, and animals

900 ***mya*** First multicellular life (underwater plants)

800 First animals (sponges)

700 Radial animals (hydra, jellyfish, comb jellies)

630 Bilateral animals (flatwormlike organisms)

600 Invertebrate protostomes (worms, arthropods, mollusks)

580 Invertebrate deuterostomes (*Pikaia,* starfish)

*543 Cambrian Explosion begins*

540 Invertebrate chordates (lancelets, urochordates)

530 Vertebrates (*Haikouella*)

505 Jawless fish (lamprey)

*490 Cambrian Explosion ends*

480 Jawed fish (most fish)

465 Plants colonize land

---

* Michael Marshall, Timeline: The Evolution of Life. *New Scientist,* July 14, 2009, www.newscientist.com/article/dn17453-timeline-the-evolution-of-life, retrieved March 17, 2017; Timeline of the Evolution of Life, Wikipedia en.wikipedia.org/wiki/Timeline_of_the_evolutionary_history_of_life; March 17, 2017.

| | |
|---|---|
| 350 | Amphibians (tetrapods with lungs) |
| 310 | Synapsids (mammal-like reptiles) |
| 305 | Sauropsids (true reptiles) |
| 230 | Dinosaurs |
| 210 | Mammals |
| 150 | Birds |
| 130 | Flowering plants |
| 70 | Primates |
| 25 | Apes |
| 6 | Humans |

***bya*** = billions of years ago

***mya*** = millions of years ago

# BIBLIOGRAPHIC KEY

Full citations for the following sources can be found online at **deep-history-of-ourselves.com**.

## Preface

Wilson (2014)

## Prologue: Why on Earth . . . ?

Dobzhansky (1973)
Emes, Grant (2001)
Keller (1973)
LeDoux (2012)
LeDoux (2014)
LeDoux (2015)
Lorenz (1965)
Skinner (1938)
Tinbergen (1951)

*Part 1: Our Place in Nature*

## Chapter 1: Deep Roots

Baluska, Mancuso (2009)
Damasio (2018)
Dennett (2017)
Emes, Grant (2011)
Gould (2001)
Grant (2016)
Jennings (1906)
Knoll (2003)
Lane (2015)
LeDoux (2012)
LeDoux (2014)
Pechère (2007)
Reber (2018)
Ryan, Grant (2009)
Tavolga (1969)
van Duijn, et al. (2006)
Wilson (2014)

## Chapter 2: The Tree of Life

Aristotle (350 BCE)
Baluska, Mancuso (2009)
Cain et al. (2007)
Darwin (1859)
Gazzaniga (2008)
Gontier (2011)
Hodos, Campbell (1969)
Lovejoy (1936)
Pollan (2002)
Smallwood et al. (1948)
Wallace (1855)
Wilson (2014)

## Chapter 3: Kingdoms Come

Beccaloni (2008)
Cavalier-Smith (2010)
Cavalier-Smith (2017)
Gould (1980)
Gould (2001)
Hagen (2012)
Lane (2015)
Margulis, Chapman (2009)
Ruggiero et al.(2015)
Scamardella (1999)
Stearns, Stearns (2000)
Steenkamp et al.(2006)
Whittaker (1957)
Woese et al.(1990)
Woese, Fox (1977)

---

https://en.wikipedia.org/wiki/Kingdom_(biology); retrieved Nov. 29, 2016.

## Chapter 4: Common Ancestry

Darwin (1859)
Dawkins (1976)
Dobzhansky (1937)
Doolittle (1999)
Fisher (1930)
Gould (1977)

Hennig (1966)
Huxley (1942)
Lane (2015)
Larson (1997)
Mayr (1974)
Mayr (1982)
Mayr (2001)
Theobald (2010)
Woese (1998)
Wright (1931)

## Chapter 5: It's a Livin' Thing

Bateson (2005)
Dawkins (1976)
Folse, Roughgarden (2010)
Grosberg, Strathmann (2007)
Jonas (1968)
Lane (2015)
Maier, Schneirla (1965)
Maturana (1975)
Michod (2005)
Niklas, Newman (2013)
Pradeu (2010)
Rokas (2008)
Torruella et al. (2015)
Varela (1997)
West, Kiers (2009)

*Part 2: Survival and Behavior*

## Chapter 6: The Behavior of Organisms

Balleine, Dickinson (1998)
Beach (1950)
Buss, Greiling (1999)
Darwin (1859)
Darwin (1872)
Darwin (1880)
Di Paolo, Thompson (2014)
Dickinson (1985)
Edmunds (1974)
Gershman, Daw (2017)
Hinde (1970)
Huxley (1942)
Jennings (1906)
LeDoux, Daw (2018)
Lehrman (1953)
Lorenz (1965)
Lyon (2015)
Maier, Schneirla (1965)
Manning (1967)
Maturana, Varela (1988)
Mayr (1963)
Morgan (1890–1891)
Niv (2007)
Pollan (2002)
Russell (1921)
Schneirla (1959)
Skinner (1938)
Smith (1993)
Staddon (1983)
Thorndike (1898)
Tinbergen (1951)
van Duijn et al. (2006)
Watson (1925)

## Chapter 7: Beyond Animal Behavior

Baluska et al. (2006)
Baluska et al. (2009)
Bengtson (2002)
Chamovitz (2013)
Darwin (1880)
Di Paolo, Thompson (2014)
di Primio et al. (2000)
Garzon (2007)
Iwatsuki, Naitoh (1988)
Jennings (1906)
Jonas (1968)
LeDoux (2012)
LeDoux (2015)
Loeb (1918)
Lorenz (1965)
Lyon (2015)
Mancuso, Viola (2015)
Maturana, Varela (1980)
Morgan (1890–1891)
Russell (1921)
Shapiro (2007)
Skinner (1938)
van Duijn et al. (2006)
Varela (1997)

## Chapter 8: The Earliest Survivors

Adler (1966)
Berg (2004)
Butler, Camilli (2005)
Fernando et al. (2009)
Greenspan (2007)
Hellingwerf (2005)
Hennessey et al. (1979)
Hoff et al. (2009)
Koonin (2003)
Koshland (1977)
LeDoux (2012)
Lee et al. (2017)
Macnab, Koshland (1972)
McGregor et al. (2012)
Moreno, Etxeberria (2005)
Pechère (2007)
Pérez-Cerezales et al. (2015)
Popkin (2017)
Ryan, Grant (2009)
Tagkopoulos et al. (2008)
Taylor, Stocker (2012)
van Duijn et al. (2006)
Vladimirov, Sourjik (2009)
Wadhams, Armitage (2004)

### Chapter 9: Survival Strategies and Tactics

Baluska, Mancuso (2009)
Bryant, Frigaard (2006)
Edmunds (1974)
Gibson et al. (2015)
Koonin (2003)
Lawrence (2002)
Mayr (1963)
Niklas (2014)
Niklas, Newman (2013)
Nilsson (1996)
Plachetzki et al. (2005)
Rittschof et al. (2014)
Roberts, Kruchten (2016)
Rokas (2008)
Schneirla (1959)
Scott-Phillips et al. (2011)
Spudich et al. (2000)
Tinbergen (1963)
Williams (2016)

### Chapter 10: Rethinking Behavior

Bengtson (2002)
Churchland (1988)
Churchland (1988)
Danziger (1997)
Darwin (1872)
Fletcher (1995)
Fletcher (1995)
Furnham (1988)
Gardner (1987)
Keller (1973)
Kelley (1992)
LeDoux (2012)
LeDoux (2015)
LeDoux (2017)
LeDoux, Brown (2017)
LeDoux, Pine (2016)
Mandler, Kessen (1964)
Marx (1951)
Michod (2005)
Romanes (1882)
Skinner (1938)
Stich (1983)
Watson (1925)

## *Part 3: Microbial Life*

### Chapter 11: In the Beginning

Alperts et al. (2002)
Cronin, Walker (2016)
Darwin (1887)
Haldane (1991)
Knoll (2003)
Lane (2015)
Lane, Le Page (2009)
Marshall (2009)
Marshall (2016)
Mastin (2009)
Maturana, Varela (1987)
Pascal, Pross (2016)
Pross (2016)
Sarafian et al. (2017)
Volk (2017)
Wickramasinghe et al. (2010)

### Chapter 12: Life Itself

Alperts et al. (2002)
Cairns-Smith (1985)
Darwin (1887)
Diemer, Stedman (2012)
Ghose (2013)
Gilbert (1986)
Hollis et al. (2000)
Holmes (2012)
Joyce (1989)
Joyce (2002)
Knoll (2003)
Lane (2009)
Lane (2015)
Lane, Le Page (2009)
Marshall (2009)
Marshall (2016)
Martin, Russell (2003)
Miller (1953)
Volk (2017)
Wachtershauser (1990)
Wachtershauser (2006)
Wickramasinghe et al. (2010)
Wikipedians (2017)

### Chapter 13: Survival Machines

Baym et al. (2016)
Cain et al. (2007)
Damper, Epstein (1981)
Gribaldo, Brochier-Armanet (2006)
Knoll (2003)
Strahl, Hamoen (2010)
Yong (2016)

### Chapter 14: The Arrival of Organelles

Cain et al. (2007)
Cavalier-Smith (2010)
Gould (2001)
Hagen (2012)
Knoll (2003)
Koonin (2003)
Lawrence (2002)
Shih, Rothfield (2006)
Whittaker (1957)
Woese et al. (1990)
Woese, Fox (1977)

## Chapter 15: The Marriage of LUCA's Children

Knoll (2003)
Lane (2015)
Margulis (1970)
Margulis, Chapman (2009)
Martin, Muller (1998)

## Chapter 16: Breathing New Life into Old

Cain et al. (2007)
Knoll (2003)
Lane (2015)
Margulis (1970)
Martin, Muller (1998)
Pollan (2002)
Volk (2017)

*Part 4: The Transition to Complexity*

## Chapter 17: Size Matters

Bengtson (2002)
Dawkins (1976)
Gerhart, Kirschner (1997)
Knoll (2003)
Lane (2014)
Lane (2015)

## Chapter 18: The Sexual Revolution

Butterfield (2000)
Crisp et al. (2015)
Dawkins (1976)
Janicke et al. (2016)
Lane (2015)
Otto (2008)
Speijer et al. (2015)
Umen, Heitman (2013)
Williams (1975)
Williams (2015)

## Chapter 19: Mitochondrial Eve, Jesse James, and the Origin of Sex

de Paula et al. (2013)
Kuijper et al. (2015)
Lane (2015)
Lane et al. (2013)
Stone et al. (2001)

## Chapter 20: Colonial Times

Bonner (1998)
Butler et al. (2010)
Folse, Roughgarden (2010)
Grosberg, Strathmann (2007)
Kirk (2005)
Lane (2015)
Lewontin (1983)
Libby, Ratcliff (2014)
Niklas (2014)
Niklas, Newman (2013)
Niklas, Newman (eds.) (2016)
Pradeu (2010)
Queller, Strassmann (2009)
Rokas (2008)
Shapiro (1998)
Waite et al. (2015)
Waters, Bassler (2005)
West, Kiers (2009)

## Chapter 21: The Selection Two-Step

Buss (1987)
Damuth, Heisler (1988)
Folse, Roughgarden (2010)
Grosberg, Strathmann (2007)
Keverne (2015)
Kirk (2005)
Lalande (1996)
Libby, Ratcliff (2014)
McGowan et al. (2009)
Michod (2005)
Nestler (2013)
Nestler et al. (2016)
Niklas (2014)
Niklas, Newman (2013)
Niklas, Newman (eds.) (2016)
Radtke et al. (2011)
Rokas (2008)
Ruiz-Trillo et al. (2008)
Smith, Szathmary (1995)

## Chapter 22: Flagellating Through the Bottleneck

Alie, Manuel (2010)
Dayel et al. (2011)
de Paula et al. (2013)
Fairclough et al. (2010)
Lapage (1925)
Levin, King (2013)
Niklas (2014)
Pettitt et al. (2002)
Reynolds, Hulsmann (2008)
Richter, King (2013)
Ruiz-Trillo et al. (2008)
Snell et al. (2001)
Umen, Heitman (2013)

---

http://www.diffen.com/difference/Cilia_vs_Flagella; retrieved Nov. 8, 2016.
http://www.hhmi.org/research/choanoflagellates-and-origin-animals; retrieved Dec. 2, 2016.

*Part 5: . . . And Then Animals Invented Neurons*

## CHAPTER 23: WHAT IS AN ANIMAL?

Boero et al. (2007)
Briggs (2013)
Cain et al. (2007)
Cavalier-Smith (2017)
Chen et al. (2004)
Collins et al. (2005)
Conway Morris (2006)
Dunn et al. (2008)
Erwin (2015)
Gold et al. (2016)
Gould (1989)
Holland (2011)
Lee et al. (2013)
Levinton (2013)
Marshall (2009)
Martindale (2005)
Moroz et al. (2014)
Rehm et al. (2011)
Schierwater et al. (2009)
Steenkamp et al. (2006)
Whelan et al. (2015)
Zapata et al. (2015)

https://en.wikipedia.org/wiki/Timeline_of_evolutionary_history_of_life; retrieved Nov. 4, 2016.

## CHAPTER 24: A HUMBLE BEGINNING

Abdul Wahab et al. (2014)
Adamska et al. (2011)
Amano, Hori (1996)
Cannon et al. (2016)
Cavalier-Smith (2017)
Erwin, Valentine (2013)
Grimaldi, Engel (2005)
Leys, Degnan (2001)
Leys, Meech (2006)
Marshall (2011)
Mukhina et al. (2006)
Nickel (2010)
Nielsen (2008)
Pennisi, Roush (1997)
Radzvilavicius et al. (2016)
Rohde et al. (2015)
Ruiz-Trillo et al. (2004)
Ruppert et al. (2003)
Schierwater et al. (2009)
Srivastava et al. (2010)
Vermeij (1996)
Yin et al. (2015)

## CHAPTER 25: ANIMALS TAKE SHAPE

Angier (2011)
Caldwell (1979)
Cavalier-Smith (2017)
Collins et al. (2005)
Dawkins (1976)
Fautin, Romano (1997)
Greenspan (2007)
Grosberg, Strathmann (2007)
Michod, Roze (2001)
Pisani et al. (2015)
Satterlie (2011)
Seipel, Schmid (2005)
Zapata et al. (2015)

http://www.encyclopedia.com/plants-and-animals/animals/zoology-invertebrates/cnidaria#3400500071; retrieved Nov. 16, 2016.

## CHAPTER 26: THE MAGIC OF NEURONS

Bear et al. (2007)
Kandel et al. (2000)
Shepherd (1983)
Sherrington (1906)

## CHAPTER 27: HOW NEURONS AND NERVOUS SYSTEMS HAPPENED

Arendt et al. (2016)
Bucher, Anderson (2015)
Conaco et al. (2012)
Dunn et al. (2008)
Elliott, Leys (2007)
Elliott, Leys (2010)
Emes, Grant (2011)
Ginsburg, Jablonka (2010)
Greenspan (2007)
Holland (2003)
Holmes (2009)
Jekely (2011)
Katsuki, Greenspan (2013)
Kelava et al. (2015)
Koizumi et al. (1990)
Kristan (2016)
Lettvin et al. (1959)
Leys (2015)
Leys, Degnan (2001)
Leys, Hill (2012)
Liebeskind (2011)
Marshall (2011)
Moroz et al. (2014)
Moroz, Kohn (2016)
Nickel (2010)
Pisani et al. (2015)
Renard et al. (2009)
Robson (2011)
Satterlie (2011)
Senatore et al. (2016)
Sherrington (1906)
Sherrington (1933)
van Duijn et al. (2006)
Whelan et al. (2015)

*Part 6: Metazoan Bread Crumbs in the Oceans*

## Chapter 28: Facing Forward

Bailly et al. (2013)
Cannon et al. (2016)
Erwin, Valentine (2013)
Finnerty (2003)
Finnerty (2005)
Finnerty et al. (2004)
Grabowsky (1994)
Greenspan (2007)
Grimaldi, Engel (2005)
Holland (2000)
Lake (1990)
Marshall (2009)
Martindale (2005)
Matus et al. (2006)
Meinhardt (2002)
Rentzsch et al. (2006)
Ruiz-Trillo et al. (2004)
Shepherd (1983)
Vermeij (1996)

---

https://www.boundless.com/biology/textbooks/boundless-biology-textbook/introduction-to-animal-diversity-27/features-used-to-classify-animals-163/animal-characterization-based-on-body-symmetry-634-11856/; retrieved Feb. 11, 2017.

## Chapter 29: Tissue Issues

Cain et al. (2007)
Cannon et al. (2016)
Chen et al. (2004)
Erwin, Davidson (2002)
Gilbert (2013)
Hejnol (2015)
Martindale et al. (2004)
Matus et al. (2006)
Northcutt (2012)
Raff (2008)
Ruiz-Trillo et al. (1999)
Ruiz-Trillo et al. (2004)
Ruppert et al. (2003)
Shepherd (1983)
Steinmetz et al. (2017)
Technau, Scholz (2003)

## Chapter 30: Oral or Anal?

Anderson (2016)
Baguna et al. (2008)
Bendesky, Bargmann (2011)
Bourlat et al. (2008)
Chen et al. (2004)
Erwin, Davidson (2002)
Fedonkin et al. (2007)
Fedonkin, Waggoner (1997)
Finnerty (2003)
Finnerty (2005)
Finnerty et al. (2004)
Gee (1996)
Gilbert (2013)
Grabowsky (1994)
Greenspan (2007)
Grimaldi, Engel (2005)
Hejnol, Martín-Durán (2015)
Hirth, Reichert (1999)
Holland (2000)
Holland (2015)
Holland et al. (2015)
Ikuta (2011)
Ivantsov (2013)
Kupfermann et al. (1991)
Martindale (2005)
Matus et al. (2006)
Meyer (1998)
Pandey, Nichols (2011)
Prince et al. (1998)
Rentzsch et al. (2006)
Ruiz-Trillo et al. (1999)
Ruiz-Trillo et al. (2004)
Ryan, Grant (2009)
Takahashi et al. (2009)
Wada, Satoh (1994)
Yin, Tully (1996)

---

https://en.wikipedia.org/wiki/Kimberella#cite_ref-Fedonkin2007_3-8; retrieved Feb. 24, 2018.

## Chapter 31: Deep-Sea Deuterostomes Link Us to Our Past

Bailly et al. (2013)
Delsuc et al. (2006)
Holland (2015)
Holland et al. (2013)
Lowe et al. (2015)
Satoh et al. (2014)

## Chapter 32: A Tale of Two Chords

Annona et al. (2015)
Bailly et al. (2013)
Bertrand, Escriva (2011)
Brunet et al. (2015)
Delsuc et al. (2006)
Hirth et al. (2003)
Holland (2015)
Holland (2015)
Holland et al. (2013)
Holland et al. (2015)
Holland, Onai (2012)
Lacalli (1994)
Lacalli (1996)
Lacalli (2001)
Lauri et al. (2014)
Lowe et al. (2015)
Mallatt, Chen (2003)
Nieuwenhuys (2002)
Putnam et al. (2008)
Satoh et al. (2014)

*Part 7: The Vertebrates Arrive*

CHAPTER 33: BAUPLAN VERTEBRATA

Arthur (1997)
Charrier et al. (2012)
Costandi (2006)
Dennis et al. (2012)
Donoghue, Purnell (2009)
Downs et al. (2008)
el-Showk (2014)
Erwin (1999)
Gould (2001)
Holland (2013)
Holland et al. (2013)
Holland et al. (2017)
Hudry et al. (2014)
Ikuta (2011)
Kumar, Hedges (1998)
Larsen (1993)
Meyer (1998)
Prince et al. (1998)
Raven, Johnson (2002)
Romer (1977)
Shubin (2008)
Valentine (2004)
Wada, Satoh (1994)
Wellik (2009)

CHAPTER 34: THE LIFE AQUATIC

Barford (2013)
Chen et al. (1999)
Downs et al. (2008)
Fouke (2017)
Gillis et al. (2009)
Gould (2001)
Grillner et al. (1998)
Helfman et al. (1997)
Long (1996)
Ota, Kuratani (2007)
Raven, Johnson (2002)
Robertson et al. (2014)
Sample (2006)
Shepherd (1983)
Shu et al. (2003)
Shubin (2008)
Zhu et al. (2012)

---

https://en.wikipedia.org/wiki/Placodermi; retrieved Jun. 17, 2017.
http://news.bbc.co.uk/1/hi/sci/tech/504776.stm; retrieved Jun. 17, 2017.
http://palaeos.com/vertebrates/placodermi/placodermi.html; retrieved Jun. 17, 2017.

CHAPTER 35: ON THE SURFACE

Benton (2001)
Bryant (2002)
Clack (2005)
Daeschler et al. (2006)
Gould (2001)
Grimaldi, Engel (2005)
Holmes (2006)
Janis (2001)
Shepherd (1983)
Shubin (2008)
Striedter (2005)

---

http://www.guinnessworldrecords.com/world-records/largest-mammal; retrieved Apr. 12, 2017.
https://www.mnn.com/earth-matters/animals/photos/11-of-the-smallest-mammals-in-the-world/etruscan-shrew; retrieved Apr. 12, 2017.

CHAPTER 36: THE MILK TRAIL

Fleagle (1999)
Janis (2001)
Kemp (2005)
Kermack, Kermack (1984)
Luo (2007)
Martin (1990)
Murray et al. (2017)
Ravosa, Dagosto (eds.) (2007)
Rowe (1988)
Striedter (2005)

*Part 8: Ladders and Trees in the Vertebrate Brain*

CHAPTER 37: NEURO-BAUPLAN VERTEBRATA

Butler, Hodos (2005)
Darwin (1859)
Darwin (1871)
Geschwind, Konopka (2012)
Holland (2015)
Holland et al. (2013)
Holland et al. (2017)
Nauta, Karten (1970)
Nieuwenhuys (2002)
Preuss et al. (2004)
Shepherd (1983)
Sprecher, Reichert (2003)
Striedter (2005)
Wada, Satoh (1994)

CHAPTER 38: LUDWIG'S LADDER

Ariens Kappers (1921)
Butler, Hodos (2005)
Edinger, Rand (1908)
Grillner et al. (2013)
Gunturkun, Bugnyar (2016)
Herrick (1948)

Hodos, Campbell (1969)
Kaas (1995)
Kaas (2011)
Karten (1991)
Karten (2015)
Karten, Shimizu (1989)
Krubitzer, Kaas (2005)
Lanuza et al. (1998)
Martinez-Garcia et al. (2002)
Nauta, Karten (1970)
Northcutt (1981)
Northcutt (2002)
Northcutt (2012)
Northcutt, Kaas (1995)
Pabba (2013)
Papez (1937)
Preuss (2012)
Reiner (1990)
Reiner (2009)
Reiner et al. (1998)
Shepherd (1983)
Smulders (2009)
Striedter (2005)
Swanson (1983)


## Chapter 39: The Triune Temptress


Brodal (1982)
Butler, Hodos (2005)
Edinger, Rand (1908)
Kluver, Bucy (1937)
Kotter, Meyer (1992)
LeDoux (1987)
LeDoux (1991)
LeDoux (1996)
LeDoux (2015)
MacLean (1949)
MacLean (1952)
MacLean (1970)
MacLean (1990)
Panksepp (1980)
Panksepp (1998)
Panksepp (2011)
Panksepp (2016)
Panksepp, Biven (2012)
Papez (1937)
Reiner (1990)
Sagan (1977)
Striedter (2005)
Swanson (1983)


## Chapter 40: Darwin's Muddled Emotional Psychology


Darwin (1872)
Descartes (1637)
Keller (1973)
Kennedy (1992)
Knoll (1997)
Mitchell et al. (eds.)
Morgan (1930)
Morris (1967)
Penn et al. (2008)


## Chapter 41: How Basic Are Basic Emotions?


Anderson, Phelps (2002)
Bertini et al. (2013)
Cannon (1929)
Coan (2010)
de Waal (2016)
Ekman (1993)
Feinstein et al. (2013)
Hess (1962)
Hoppenbrouwers et al. (2016)
Izard (1971)
Izard (1990)
James (1884)
James (1890)
LeDoux (1996)
LeDoux (2012)
LeDoux (2014)
LeDoux (2015)
LeDoux et al. (2018)
LeDoux, Pine (2016)
MacLean (1949)
MacLean (1952)
MacLean (1970)
MacLean (1990)
Öhman (2005)
Olsson, Phelps (2004)
Panksepp (1998)
Panksepp (2005)
Panksepp (2011)
Papez (1937)
Plutchik (1980)
Scarantino (2018)
Tamietto and de Gelder (2010)
Tamietto et al. (2009)
Tamietto et al. (2012)
Tomkins (1962)
Tomkins (1963)


*Part 9: The Beginning of Cognition*

## Chapter 42: Cogitation


Bargh (1997)
Boakes (1984)
Boring (1950)
Cerullo (2015)
Chamovitz (2013)
Darwin (1872)
Dehaene et al. (2017)
Descartes (1637)
Freud (1915)
Gardner (1987)
Hassin et al. (eds.) (2005)
James (1890)
Keller (1973)
Kennedy (1992)
Kihlstrom (1987)
Kurzweil (1999)
Lashley (1958)
Lorenz (1950)
Mancuso, Viola (2015)
Mitchell et al. (eds.) (1996)
Radman (ed.) (2017)

Reber (2018)
Richards (ed.) (2001)
Ryle (1949)
Skinner (1938)
Terrace, Metcalfe (2004)
Tinbergen (1951)
Tononi et al. (2016)
van Duijn et al. (2006)
Watson (1925)

https://plato.stanford.edu/entries/cognitive-science/; retrieved Sept. 14, 2017.

## Chapter 43: Finding Cognition in the Behaviorist Bailiwick

Avargues-Weber et al. (2012)
Balleine, Dickinson (1998)
Boakes (1984)
Buckner (2011)
Byrne, Bates (2006)
Cheeseman et al. (2014)
Cheung et al. (2014)
Clayton et al. (2001)
Colwill, Rescorla (1990)
Daw (2015)
Dayan (2008)
Decker et al. (2016)
Dickinson (1985)
Dickinson (2012)
Dolan, Dayan (2013)
Emes, Grant (2011)
Garcia et al. (1955)
Giurfa (2012)
Giurfa et al. (2001)
Glanzman (2010)
Gould (2004)
Grant (2016)
Gunturkun, Bugnyar (2016)
Hawkins, Byrne (2015)
Heisenberg (2015)
Hinde (1970)
Holland (1993)
Holland, Rescorla (1975)
Kandel (2001)
Lechner, Byrne (1998)
Maier, Schneirla (1965)
McCurdy et al. (2013)
Minors (2016)
Murray et al. (2017)
Muzio et al. (2011)
O'Keefe, Dostrovsky (1971)
O'Keefe, Nadel (1978)
Papini (2010)
Pavlov (1927)
Perry et al. (2013)
Pickens, Holland (2004)
Roberts, Glanzman (2003)
Seligman, Hager (eds.) (1972)
Skinner (1938)
Sorabji (1993)
Thorndike (1898)
Thorndike (1905)
Tolman (1932)
Tolman (1948)
Watson (1925)
Wilkinson, Huber (2012)
Wynne, Udell (2013)

https://plato.stanford.edu/entries/cognition-animal/; retrieved Aug 2, 2017.

## Chapter 44: The Evolution of Behavioral Flexibility

Averbeck, Costa (2017)
Balleine, Dickinson (1998)
Berridge (2007)
Berridge, Kringelbach (2015)
Cardinal et al. (2002)
Clayton, Dickinson (1998)
Correia et al. (2007)
Darwin (1872)
Daw (2014)
Dayan, Watkins (2006)
Everitt, Robbins (2005)
Feeney et al. (2009)
Glimcher (2011)
Gunturkun, Bugnyar (2016)
Hamid et al. (2016)
Hart et al. (2014)
Lattal (1998)
MacLean (1949)
MacLean (1952)
MacLean (1970)
MacLean (1990)
Murray et al. (2017)
Muzio et al. (2011)
Niv et al. (2005)
Olds (1956)
Olds, Milner (1954)
Panksepp (1980)
Panksepp (1998)
Papini (2010)
Reynolds, Wickens (2002)
Romanes (1882)
Schultz et al. (1997)
Skov-Rackette et al. (2006)
Thorndike (1905)
Ward-Fear et al. (2016)
Wise (1980)
Zentall et al. (2001)
Zinkivskay et al. (2009)

*Part 10: Surviving (and Thriving) by Thinking*

## Chapter 45: Deliberation

Baddeley (2003)
Baum (2003)
Beran et al. (2016)
Doll et al. (2015)
Gillan et al. (2015)
Holyoak (2005)
Johnson-Laird (1983)
Johnson-Laird (2006)
Kahneman (2011)
Levitin (2015)
MacLean (2016)
Murray et al. (2017)

O'Keefe, Nadel (1978)
Otto et al. (2013)
Otto et al. (2015)
Penn et al. (2008)
Pinker (1997)
Raby, Clayton (2009)
Schneider, Shiffrin (1977)
Simon, Daw (2011)
Tolman (1948)
Tomasello, Rakoczy (2003)

## CHAPTER 46: THE ENGINE OF DELIBERATIVE COGNITION

Alexander (2016)
Baddeley (1986)
Baddeley (1992)
Bartlett (1923)
Bartlett (1932)
Beck (1976)
Binder, Desai (2011)
Bowlby (1969)
Cowan (1988)
Cowan (2001)
Curtis, Lee (2010)
Custers, Aarts (2010)
Daw et al. (2005)
Daw et al. (2006)
Dehaene et al. (2017)
D'Esposito, Postle (2015)
Eriksson et al. (2015)
Fan (2014)
Fuster (2008)
Goldman-Rakic (1996)
Graham et al. (2010)
Hayes (2019)
Heyes (2016)
Horner, Burgess (2014)
Hunsaker, Kesner (2013)
Javanbakht (2011)
Johnson-Laird (2010)
Kiefer (2012)
Kim (2016)
Koechlin (2011)
Lambon Ralph (2014)
Lau, Passingham (2007)
LeDoux, Daw (2018)
Luck, Vogel (1997)
Ma et al. (2014)
Mandler (1984)
Marr (1971)
Mattson (2014)
McCurdy et al. (2013)
Miller (1956)
Miller (2013)
Miller, Cohen (2001)
Minsky (1975)
Murray et al. (2017)
Murray et al. (2017)
O'Reilly, McClelland (1994)
Otto et al. (2013)
Penn et al. (2008)
Piaget (1929)
Pidgeon, Morcom (2016)
Postle (2006)
Price et al. (2015)
Rolls (2013)
Rumelhart (1980)
Smith et al. (2012)
Smith et al. (2014)
Soto, Silvanto (2014)
Thompson et al. (2015)
Trubutschek et al. (2017)
Valadao et al. (2015)
Young et al. (2003)

## CHAPTER 47: SCHMOOZING

Bowerman, Levinson (eds.) (2001)
Carruthers (2008)
Carruthers, Ritchie (2012)
Chomsky (1973)
Corballis (2017)
Dennett (1991)
Dennett (1996)
Dunbar (1998)
Everett (2012)
Fodor (1975)
Godfrey-Smith (2016)
Gould (1991)
Gould (1997)
Gould (2007)
Gould, Lewontin (1979)
Gould, Vrba (1982)
Harari (2015)
Hayes (2019)
Herrmann et al. (2007)
Heyes (2018)
Hoffmann et al. (2018)
Hoffmann et al. (2018)
Horner, Burgess (2014)
Javanbakht (2011)
Kitayama, Markus (eds.) (1994)
Koerner (2000)
Kolodny, Edelman (2018)
Lakoff (1987)
MacLean (2016)
Mattson (2014)
Penn et al. (2008)
Pidgeon, Morcom (2016)
Pinker (1994)
Preuss (2011)
Rolls (2008)
Seidner (1982)
Shatz (2008)
Shea et al. (2014)
Tomasello, Rakoczy (2003)
Vygotsky (1934)
Weiskrantz (1997)
Whorf (1956)
Wierzbicka (1994)
Wittgenstein (1958)
Wolff, Holmes (2011)

*Part 11: Cognitive Hardware*

## Chapter 48: Perception and Memory Share Circuitry


Amaral (1987)
Binder, Desai (2011)
Catani et al. (2005)
Clarke et al. (2013)
Damasio (1989)
DiCarlo et al. (2012)
Eichenbaum (2017)
Eichenbaum (2017)
Felleman, Van Essen (1991)
Friederici (2017)
Fuster (2008)
Gauthier et al. (2003)
Goldman-Rakic (1987)
Goldman-Rakic (1996)
Graham et al. (2010)
Gross (1994)
Hagoort (2014)
Hubel (1988)
Kondo et al. (2005)
Lambon Ralph (2014)
Livingstone (2008)
McCurdy et al. (2013)
Mesulam (1998)
Milner, Goodale (2006)
Mishkin et al. (1983)
Miyashita (1993)
Murray et al. (2017)
Rademaker et al. (2018)
Rilling et al. (2011)
Ritchey et al. (2015)
Rolls (2000)
Schiller, Tehovnik (2015)
Seltzer, Pandya (1978)
Squire (1987)
Thompson et al. (2015)
Tulving (1972)
Ungerleider, Mishkin (1982)
Wang, Morris (2010)
Yeterian et al. (2012)
Young (1992)


## Chapter 49: The Cognitive Coalition


Amaral (1987)
Badre, D'Esposito (2009)
Bar (2003)
Barbas et al. (1999)
Barbas, García-Cabezas (2016)
Barbas, Pandya (1989)
Bergstrom, Eriksson (2014)
Berryhill et al. (2011)
Bettcher et al. (2016)
Binder, Desai (2011)
Burgess, Stuss (2017)
Cabeza, St. Jacques (2007)
Carlen (2017)
Carter et al. (1998)
Christophel et al. (2017)
Clarke, Tyler (2015)
Craig (2003)
Craig (2009)
Damasio (1989)
Damasio (1999)
Daw et al. (2006)
D'Esposito, Postle (2015)
Eichenbaum (2017)
Eriksson et al. (2015)
Fan (2014)
Fleming et al. (2014)
Fuster (2008)
Gauthier et al. (2003)
Goldman-Rakic (1987)
Goldman-Rakic (1996)
Horner, Burgess (2014)
Javanbakht (2011)
Joyce, Barbas (2018)
Koechlin et al. (1999)
Koechlin et al. (2003)
Koechlin, Hyafil (2007)
Koechlin, Summerfield (2007)
Kondo et al. (2005)
Krawczyk (2012)
Kringelbach (2005)
Lambon Ralph (2014)
Lara, Wallis (2015)
Lau, Passingham (2007)
Lewis et al. (2002)
Lewis-Peacock, Postle (2008)
Libby et al. (2014)
Liu et al. (2013)
Mattson (2014)
McCurdy et al. (2013)
Mesulam (1998)
Miller, Cohen (2001)
Moayedi et al. (2015)
Neubert et al. (2014)
Okuda et al. (2003)
Ongur et al. (2003)
Otto et al. (2013)
Passingham (1995)
Passingham, Wise (2012)
Petrides et al. (2012)
Petrides, Pandya (1988)
Pezzulo et al. (2018)
Pidgeon, Morcom (2016)
Posner, DiGiralomo (1998)
Postle (2006)
Postle (2016)
Rademaker et al. (2018)
Rahnev (2017)
Ramnani, Owen (2004)
Ritchey et al. (2015)
Rolls (2014)
Romanski (2004)
Rushworth et al. (2007)
Seltzer, Pandya (1978)
Sreenivasan et al. (2014)
Thompson et al. (2015)
Wang, Morris (2010)
Wise (2008)
Yeterian et al. (2012)
Young (1992)
Zanto et al. (2011)

## Chapter 50: Rewired and Running Hot


Allman et al. (2010)
Barbas, García-Cabezas (2016)
Bastos et al. (2018)
Boorman et al. (2009)
Carlen (2017)
Damasio (1994)
Damasio et al. (1994)
Donahue et al. (2018)
Finlay et al. (1998)
Friederici (2017)
Fuster (2008)
Hagoort (2014)
Jerison (1973)
Joyce, Barbas (2018)
Kaas (1995)
Kaas (2011)
Koechlin (2011)
Koechlin et al. (1999)
Koechlin et al. (2003)
Koechlin, Hyafil (2007)
Koechlin, Summerfield (2007)
Konopka et al. (2012)
Krubitzer, Kaas (2005)
LeDoux, Brown (2017)
Luria (1973)
Moayedi et al. (2015)
Murray et al. (2017)
Neubert et al. (2014)
Nimchinsky et al. (1999)
Northcutt, Kaas (1995)
Ongur et al. (2003)
Passingham (1995)
Passingham, Wise (2012)
Petrides et al. (2012)
Preuss (1995)
Preuss (2011)
Preuss (2012)
Rilling et al. (2008)
Rilling et al. (2011)
Schenker et al. (2008)
Semendeferi et al. (2011)
Teffer, Semendeferi (2012)
Uylings et al. (2003)
Wise (2008)


*Part 12: Subjectivity*

## Chapter 51: Being There


Benney, Henkel (2006)
Cohen, Squire (1980)
Corkin (1968)
Festinger (1957)
Gardner (1987)
Gazzaniga (1970)
Gazzaniga (1985)
Gazzaniga (2008)
Gazzaniga et al. (1962)
Gazzaniga, LeDoux (1978)
Graf, Schacter (1985)
Jarcho et al. (2011)
Johnson (2006)
LeDoux (1996)
LeDoux (2017)
Milner (1959)
Nisbett, Wilson (1977)
Noë (2012)
Pinto et al. (2017)
Radman (ed.) (2017)
Squire (1986)
Tulving (1983)


## Chapter 52: What Is It Like to Be Conscious?


Baars (1988)
Baars, Franklin (2007)
Baddeley (2000)
Baddeley, Hitch (1974)
Banaji, Greenwald (2013)
Block (2007)
Block et al. (2014)
Breitmeyer, Ogmen (2006)
Brown (2015)
Carrasco (2011)
Chalmers (1996)
Crick, Koch (1995)
Damasio (1989)
Damasio (1999)
Dehaene (2014)
Dehaene et al. (2006)
Dennett (1991)
Edelman, Tononi (2000)
Ericsson, Simon (1993)
Festinger (1957)
Freud (1915)
Friston (2013)
Frith (2007)
Frith (2008)
Frith et al. (1999)
Frith, Dolan (1996)
Gallagher, Zahavi (2012)
Gardner (1987)
Gazzaniga (1970)
Gazzaniga (1985)
Gazzaniga (2008)
Gazzaniga (2015)
Gazzaniga (2018)
Gazzaniga, LeDoux (1978)
Gilboa et al. (2006)
Graziano (2013)
Hameroff, Penrose (2014)
Jack, Shallice (2001)
Jarcho et al. (2011)
Johnson-Laird (1988)
Kahneman (1999)
Lau, Passingham (2006)
Lazarus, McCleary (1951)
LeDoux (2017)
Maier (1931)
Maniscalco, Lau (2016)
Milner (1959)
Moore (1988)
Moscovitch (1992)
Nagel (1974)
Neisser (1967)
Nisbett, Wilson (1977)
Norman, Shallice (1986)
Öhman (2002)
Öhman (2005)
Overgaard, Sandberg (2014)
Packard (1957)

Penrose (1994)
Posner (1994)
Prinz (2012)
Radman (ed.) (2017)
Robinson, Clore (2002)
Rosenthal (2005)
Schacter (1990)
Seth et al. (2008)
Shallice (1988)
Suddendorf, Redshaw (2017)
Tononi et al. (2016)
Tononi, Koch (2015)
Tulving (1983)
Tweedy (2018)
Wilson (1994)
Yang et al. (2014)

## CHAPTER 53: I WANT TO TAKE YOU HIGHER

Baars (1988)
Baars, Franklin (2007)
Block (2007)
Block et al. (2014)
Brown (2015)
Carruthers (2000)
Cleeremans (2008)
Cleeremans (2011)
Cooney, Gazzaniga (2003)
Crick, Koch (1995)
Dehaene (2014)
Dehaene, Changeux (2011)
Gennaro (2011)
Giles et al. (2016)
Gottlieb (2017)
Kriegel (2009)
Lau, Brown (in press)
Lau, Rosenthal (2011)
Mashour (2018)
McGovern, Baars (2007)
Metzinger (2003)
Naccache (2018)
Rosenthal (2004)
Rosenthal (2005)
Rosenthal (2012)
Rosenthal (2012)
Rosenthal, Weisberg (2008)
Weisberg (2011)

## CHAPTER 54: HIGHER AWARENESS IN THE BRAIN

Block (2011)
Block (2014)
Brown (2014)
Brown (2015)
Cohen et al. (2016)
Fleming et al. (2010)
Fleming et al. (2018)
Fleming, Lau (2014)
Frith (2007)
Haun et al. (2017)
Koechlin et al. (1999)
Koechlin, Hyafil (2007)
Lau, Brown (in press)
Lau, Passingham (2007)
Lau, Rosenthal (2011)
Lau, Rosenthal (2011)
LeDoux, Brown (2017)
Liu et al. (2013)
Odegaard et al. (2017)
Odegaard et al. (2018)
Ramnani, Owen (2004)
Rosenthal (2005)
Rosenthal (2012)
Shekhar, Rahnev (2018)
Sperling (1960)
Tsuchiya et al. (2015)

*Part 13: Consciousness Through the Looking Glass of Memory*

## CHAPTER 55: THE INVENTION OF EXPERIENCE

Alexander, Brown (2018)
Allport (1955)
Bar et al. (2006)
Barrett (2017)
Binder, Desai (2011)
Bruner, Goodman (1947)
Bruner, Minturn (1955)
Bruner, Postman (1949)
Cavanagh (2011)
Clark (1998)
Clark, Chalmers (1998)
Clarke et al. (2013)
Edelman (2004)
Edelman, Tononi (2000)
Fletcher, Frith (2009)
Friston (2013)
Friston, Frith (2015)
Frith (2007)
Gregory (1974)
Gregory (1997)
Grossberg (1980)
Hering (1870)
James (1890)
Lamme (2015)
Lamme (2015)
Melloni (2015)
Meyer (2011)
Neri (2014)
Panichello et al. (2012)
Pezzulo et al. (2018)
Rao, Ballard (1999)
Rensink, O'Regan (1997)
Schmack et al. (2013)
Seth (2016)
Seth (in press)
Seth et al. (2008)
Seth, Friston (2016)
Simons, Chabris (1999)
Simons, Levin (1997)
Sperling (1960)
Stefanics et al. (2014)
Thompson, Madigan (2005)
von Foerster (1984)
von Helmholtz (1866)
von Helmholtz (2005)
Ye et al. (2018)

## Chapter 56: Ah, Memory


Baker (2000)
Baker (2013)
Baker (2013)
Banaji, Greenwald (2013)
Binder, Desai (2011)
Chun, Jiang (2003)
Chun, Phelps (1999)
Clarke et al. (2013)
Cohen, Squire (1980)
Conway (2009)
Conway, Pleydell-Pearce (2000)
Damasio (1999)
Damasio (2010)
Dennett (1992)
Forgione (2018)
Gallagher (2000)
Henke (2010)
James (1890)
Klein (2004)
Lewis (2011)
Lewis (2011)
Lewis (2011)
Lewis (2013)
Lewis (2013)
Loaiza, Borovanska (2018)
Markowitsch, Staniloiu (2011)
Markus, Kitayama (1991)
Menant (2006)
Metcalfe, Son (2012)
Metzinger (2003)
Milner (1959)
Milner (1962)
Milner et al. (1968)
Moscovitch (1992)
Neisser (ed.) (1993)
Pasquali et al. (2010)
Roberts, Feeney (2009)
Rosenthal (2012)
Schacter, Tulving (1982)
Shea et al. (2014)
Smith (2017)
Squire (1987)
St. Jacques et al. (2018)
Sui, Humphreys (2015)
Tulving (1972)
Tulving (1983)
Tulving (2005)
Varela et al. (1993)
Waidergoren et al. (2012)
Wheeler et al. (1997)
Wilson (2002)


## Chapter 57: Putting Memories in Their Places


Amaral (1987)
Bar (2003)
Barbas et al. (1999)
Barbas, Pandya (1989)
Bertossi et al. (2016)
Binder et al. (2016)
Binder, Desai (2011)
Botzung et al. (2008)
Buckner, Carroll (2007)
Burgess (2014)
Burgess, O'Keefe (2011)
Buzsaki (2011)
Cabeza, Moscovitch (2013)
Cabeza, St. Jacques (2007)
Catani et al. (2005)
Chan et al. (2011)
Clarke et al. (2013)
Cohen, Squire (1980)
Craig (2009)
Damasio (1989)
Damasio (2010)
Denny et al. (2012)
DiCarlo et al. (2012)
Ding et al. (2009)
Dudai, Morris (2013)
Eichenbaum (2017)
Eichenbaum (2017)
Eichenbaum (2017)
Fan et al. (2014)
Feinberg (2001)
Fossati (2013)
Frith, Happé (1999)
Fuster (2008)
Gilboa (2004)
Goldman-Rakic (1987)
Goldman-Rakic (1996)
Graham et al. (2010)
Gross (1994)
Hirstein (2011)
Johnson et al. (2006)
Joyce, Barbas (2018)
Kalenzaga et al. (2014)
Kim (2012)
Kim (2016)
Klein (2004)
Koechlin et al. (1999)
Koechlin et al. (2003)
Koechlin, Hyafil (2007)
Koechlin, Summerfield (2007)
Kondo et al. (2003)
Kondo et al. (2005)
Koshino et al. (2014)
Lambon Ralph (2014)
LeDoux (2015)
Levine (2004)
Levine et al. (2004)
Lewis-Peacock, Postle (2008)
Libby et al. (2014)
Long et al. (2016)
Markowitsch, Staniloiu (2011)
Martin et al. (1995)
Martinelli et al. (2013)
McClelland et al. (1995)
McCormick et al. (2017)
McCurdy et al. (2013)
Mesulam (1998)
Milner (1962)
Milner et al. (1968)
Mishkin (1982)
Miyashita (1993)
Moscovitch (1995)
Moscovitch et al. (2005)
Moscovitch, Winocur (2002)
Murray (1992)
Murray et al. (2017)
Neubert et al. (2014)
O'Keefe, Nadel (1978)
O'Reilly, McClelland (1994)
Passingham (1995)
Passingham, Wise (2012)
Petrides et al. (2012)
Petrides, Pandya (1988)
Price et al. (2015)
Ramnani, Owen (2004)
Richter et al. (2016)
Ritchey et al. (2015)
Rolls (2000)

Schacter, Tulving (1982)
Seltzer, Pandya (1978)
Shea et al. (2014)
Squire (1987)
Squire, Zola (1998)
Sui, Humphreys (2015)
Suzuki, Amaral (2003)
Teyler, DiScenna (1986)
Thompson et al. (2015)
Tulving (1972)
Tulving (1983)
Tulving (1983)
Tulving (2005)
Uddin (2011)
Ungerleider, Mishkin (1982)
van der Meer et al. (2010)
Wang, Morris (2010)
Warrington, Shallice (1984)
Wheeler et al. (1997)
Yeterian et al. (2012)

## Chapter 58: Higher-Order Awareness Through the Lens of Memory

Baker (2013)
Clark (1998)
Cleeremans (2008)
Cleeremans (2011)
Craig (2009)
Damasio (2010)
Forgione (2018)
Friston (2013)
Friston, Frith (2015)
Frith (2007)
Frith, Happé (1999)
Gallagher (2000)
Gallagher, Frith (2003)
Haun et al. (2017)
Koch (2018)
Lau, Rosenthal (2011)
Lau, Rosenthal (2011)
Lewis (2011)
Lewis (2013)
Metzinger (2003)
Odegaard et al. (2017)
Odegaard et al. (2018)
Pasquali et al. (2010)
Rosenthal (2004)
Rosenthal (2005)
Rosenthal (2012)
Seth (2016)
Seth (in press)
Sui, Humphreys (2015)

*Part 14: The Shallows*

## Chapter 59: The Tricky Problem of Other Minds

Clayton, Dickinson (2010)
Dawkins (2017)
de Waal (2006)
Dennett (1991)
Ericsson, Simon (1993)
Frith, Happé (1999)
Gray (2004)
Heyes (2015)
Heyes (2016)
Horowitz (2016)
Humphrey (1977)
Jack, Shallice (2001)
Jennings (2006)
Kahneman (1999)
Kennedy (1992)
LeDoux (2002)
LeDoux (2017)
Loftus (1996)
Maier (1931)
Mashour (2018)
Menant (2006)
Mitchell et al. (eds.) (1996)
Nahmias (2002)
Nisbett, Wilson (1977)
Overgaard, Sandberg (2014)
Radman (ed.) (2017)
Robinson, Clore (2002)
Rosenthal (2018)
Russell (1927)
Seth et al. (2008)
Shea et al. (2014)
Shettleworth (2010)
Singer (2009)
Weiskrantz (1997)
Wilson (1994)
Wynne, Bolhuis (2008)

## Chapter 60: Creeping Up on Consciousness

Baker et al. (2011)
Beran (2017)
Brandl (2016)
Bulley et al. (2017)
Butterfill, Apperly (2013)
Carruthers (2008)
Carruthers, Ritchie (2012)
Cartmill (2000)
Clayton, Dickinson (1998)
Clayton, Dickinson (2010)
Corballis (2017)
Dehaene et al. (2017)
Dennett (1991)
Ericcson, Simon (1993)
Fleming et al. (2010)
Fleming et al. (2014)
Fleming et al. (2018)
Fleming, Lau (2014)
Frith (2007)
Frith et al. (1999)
Frith, Happé (1999)
Gallagher, Frith (2003)
Gallup (1982)
Gallup et al. (2014)
Gray (2004)
Güntürkün, Bugnyar (2016)
Heyes (1995)
Heyes (2008)
Heyes (2015)
Heyes (2016)
Heyes (2017)
Horner, Burgess (2014)
Horowitz (2016)
Jack, Shallice (2001)
Jackendoff (2007)
LeDoux (2015)
LeDoux (2017)
LeDoux, Brown (2017)
Maniscalco, Lau (2016)

Metcalfe, Son (2012)
Morales et al. (2018)
Murray et al. (2017)
Naccache, Dehaene (2007)
Nahmias (2002)
Nisbett, Wilson (1977)
Overgaard, Sandberg (2014)
Pepperberg (2009)
Peters et al. (2017)
Peters et al. (2017)
Povinelli, Prince (1998)
Premack, Woodruff (1978)
Raby, Clayton (2009)
Radman (ed.) (2017)
Redshaw et al. (2017)
Reiss, Marino (2001)
Ruby et al. (2018)
Salwiczek et al. (2010)
Schneider et al. (2017)
Seth et al. (2008)
Shea et al. (2014)
Shettleworth (2010)
Smith et al. (2012)
Suddendorf, Butler (2014)
Suddendorf, Corballis (2007)
Suddendorf, Corballis (2010)
Suddendorf, Redshaw (2017)
Terrace, Metcalfe (2004)
Tulving (1972)
Tulving (2001)
Tulving (2005)
Weiskrantz (1997)

## Chapter 61: Kinds of Minds

Beran et al. (2016)
Cartmill (2000)
Clayton, Dickinson (1998)
Clayton, Dickinson (2010)
Dennett (1996)
Güntürkün, Bugnyar (2016)
Hassin et al. (2009)
Jacob et al. (2015)
Jacobs, Silvanto (2015)
Joglekar et al. (2018)
Lau, Passingham (2007)
Mashour (2018)
Menant (2006)
Panagiotaropoulos et al. (2012)
Rosenthal (2012)
Soon et al. (2008)
Soto et al. (2011)
Soto, Silvanto (2014)
Tulving (2005)
van Vugt et al. (2018)
Wittgenstein (1958)
Wynne (2004)
Wynne (2007)

*Part 15: Emotional Subjectivity*

## Chapter 62: The Slippery Slopes of Emotional Semantics

Anderson (2016)
Anderson, Adolphs (2014)
Bacon (1620)
Bekoff (2000)
Berridge, Kringelbach (2015)
Bertini et al. (2013)
Block (1995)
Burghardt (1991)
Darwin (1872)
Dawkins (2012)
Dawkins (2017)
de Waal (1999)
Dehaene et al. (2017)
Descartes (1637)
Ekman (1993)
Fanselow, Pennington (2017)
Fanselow, Pennington (2018)
Gibson et al. (2015)
Griffin (1976)
Griffin (2015)
Hess (1962)
Heyes (1995)
Heyes (2008)
Heyes (2015)
Heyes (2016)
Heyes (2017)
Huxley (1954)
Izard (1990)
Keller (1973)
Kennedy (1992)
Lang (1968)
LeDoux (1984)
LeDoux (1987)
LeDoux (1994)
LeDoux (1996)
LeDoux (2012)
LeDoux (2013)
LeDoux (2017)
MacLean (1949)
MacLean (1952)
MacLean (1970)
Mandler, Kessen (1964)
Marx (1951)
Mitchell et al. (eds.) (1996)
Olds (1956)
Panksepp (1998)
Panksepp, Biven (2012)
Papez (1937)
Penn et al. (2008)
Penn, Povinelli (2007)
Penn, Povinelli (2007)
Perusini, Fanselow (2015)
Povinelli (2008)
Povinelli, Preuss (1995)
Povinelli, Prince (1998)
Rachman, Hodgson (1974)
Rivas, Burghardt (2002)
Romanes (1882)
Rosen, Schulkin (1998)
Scarantino (2018)
Schacter (1987)
Shettleworth (2009)
Shettleworth (2010)
Squire (1987)
Sternson et al. (2013)
Suddendorf, Corballis (2010)
Tamietto and de Gelder (2010)
Tamietto et al. (2009)
Tamietto et al. (2012)
Timberlake (1999)
Tomkins (1962)
Tomkins (1963)
Tononi, Koch (2015)
Viegas (2015)

Wise (1980)
Wynne (2004)
Wynne (2007)
Wynne, Bolhuis (2008)
Wynne, Udell (2013)


## Chapter 63: Can Survival Circuits Save the Day?


Adolphs (2013)
Adolphs, Anderson (2018)
Anderson (2016)
Bowlby (1969)
Bucci (1997)
Damasio (1994)
Damasio (1999)
Damasio, Carvalho (2013)
Doyle, Csete (2011)
Fanselow, Pennington (2017)
Fanselow, Pennington (2018)
Hoppenbrouwers et al. (2016)
Leahy (2015)
LeDoux (2012)
LeDoux (2013)
LeDoux (2014)
LeDoux (2015)
LeDoux (2017)
LeDoux, Hofmann (2018)
LeDoux, Mobbs (2018)
LeDoux, Pine (2016)
Mobbs (2018)
Mobbs et al. (2015)
Pavlov (1936)
Panksepp (1998)
Panksepp, Biven (2012)
Petrovich et al. (2001)
Sternson (2013)


## Chapter 64: Thoughtful Feelings


Alberini and LeDoux (2013)
Allbritton (1995)
Barrett (2017)
Barrett (2017)
Barrett et al. (2007)
Barrett et al. (2007)
Barrett et al. (eds.) (2007)
Barrett, Bar (2009)
Barrett, Russell (eds.) (2015)
Beck (1976)
Beck, Haigh (2014)
Brosch et al. (2013)
Cleeremans (2008)
Cleeremans (2011)
Clore, Ortony (2013)
Coan, Gonzalez (2015)
Craig (2003)
Craig (2009)
Critchley et al. (2004)
Damasio, Carvalho (2013)
Daneshmandi et al. (2018)
de Sousa (2013)
Fehr, Russell (1984)
Frijda (1986)
Hofmann (2016)
Hofmann, Doan (2018)
Kahneman (1999)
Kövecses (2000)
Kron et al. (2010)
Lazarus (1991)
Leahy (2015)
LeDoux (1984)
LeDoux (1996)
LeDoux (2012)
LeDoux (2014)
LeDoux (2017)
LeDoux, Brown (2017)
LeDoux, Hofmann (2018)
Lindquist et al. (2006)
Lindquist et al. (2015)
Lindquist, Barrett (2008)
McNally (2009)
Miloyan, Suddendorf (2015)
Nader and Einarsson (2010)
Oatley, Johnson-Laird (2014)
Ochsner, Gross (2005)
Oosterwijk et al. (2015)
Ortony, Turner (1990)
Pessoa (2013)
Robinson, Clore (2002)
Russell (2003)
Russell (2014)
Satpute et al. (2016)
Saxe, Houlihan (2017)
Schachter, Singer (1962)
Scherer (1984)
Scherer (2000)
Sloan et al. (2018)
Tulving (2005)
Wilson-Mendenhall et al. (2011)
Young et al. (2003)


## Chapter 65: Emotional Brains Run HOT


Anderson, Phelps (2002)
Barrett, Bar (2009)
Barrett, Simmons (2015)
Bertini et al. (2013)
Burnemann et al. (2012)
Brandl et al. (2017)
Brown (2015)
Bulley et al. (2017)
Cardinal et al. (2002)
Craig (2003)
Craig (2009)
Critchley et al. (2004)
Damasio (1994)
Damasio (1999)
Damasio, Carvalho (2013)
Everitt, Robbins (2005)
Feinstein et al. (2013)
Furl et al. (2013)
Ghaziri et al. (2017)
Gu et al. (2013)
Hofmann (2016)
Hofmann, Doan (2018)
Hoppenbrouwers et al. (2016)
Kron et al. (2010)
Lau, Brown (2019)
Lau, Rosenthal (2011)
LeDoux (1996)
LeDoux (2002)
LeDoux (2008)
LeDoux (2012)

LeDoux (2015)
LeDoux et al. (2018)
LeDoux, Brown (2017)
LeDoux, Daw (2018)
LeDoux, Hofmann (2018)
LeDoux, Pine (2016)
Lindquist et al. (2015)
Miloyan, Suddendorf (2015)
Murray et al. (2017)
Pessoa (2013)
Pezzulo et al. (2015)
Phelps (2006).
Satpute et al. (2013)
Saxe, Houlihan (2017)
Seth, Friston (2016)
Seymour, Dolan (2008)
Shiflett, Balleine (2010)
Sontheimer et al. (2017)
Tamietto and de Gelder (2010)
Tamietto et al. (2009)
Tamietto et al. (2012)
Wilson-Mendenhall et al. (2011)

## CHAPTER 66: SURVIVAL IS DEEP, BUT OUR EMOTIONS ARE SHALLOW

Buss (1995)
Buss et al. (1998)
Dennett (1996)
Dugatkin, Trut (2017)
Gould (1991)
Gould (1997)
Gould, Lewontin (1979)
Gould, Vrba (1982)
Hayes (2019)
Kolodny, Edelman (2018)
Menant (2011)
Miloyan, Suddendorf (2015)
Pinker (1997)
Tooby, Cosmides (1992)

## EPILOGUE: CAN WE SURVIVE OUR SELF-CONSCIOUS SELVES?

Andriessen (2006)
Balter (2015)
Dawkins (1976)
Doyle, Csete (2011)
Durkheim (1951)
Frank (2018)
Gazzaniga (2012)
Harari (2015)
Hayes (2019)
Hepburn (2018)
Huxley (1954)
Jonas (1968)
Jones (1986)
Lane (2015)
Lau (2017)
Maturana, Varela (1980)
May (2018)
Menant (2018)
Michod (2005)
Niklas, Newman (2013)
Pena-Guzman (2017)
Varela (1996)
Varela (1997)
Volk (2008)
Volk (2017)

# ILLUSTRATION CREDITS

Figure 2.1: [Public domain] from E. Haeckel (1874), *Anthropogenie oder Entwickelungsgeschichte des Menschen. Gemeinverständliche wissenschaftliche Vorträge über die Grundzüge der menschlichen Keimes- und Stammes-Geschichte* (Leipzig: Engelmann); scanned from *Tafel XII* by Hanno in 2002.

Figure 8.1: Format of illustration is based on figure 1 from M. van Duijn, F. Keijzer, D. Franken (2006), "Principles of Minimal Cognition: Casting Cognition as Sensorimotor Coordination." *Adaptive Behavior* 14:157–70.

Figure 11.1: Format based on *Stargazing Live,* a BBC and Open University coproduction: https://bit.ly/1NNUGqt.

Figure 12.2: Format based on figure 1 in G. F. Joyce (1989), "RNA Evolution and the Origins of Life," *Nature* 338:217–24.

Figure 12.3: Format of the left image is based on U.S. federal government work, public domain: https://oceanexplorer.noaa.gov/explorations/02fire/background/hirez/chemistry-hires.jpg; https://commons.wikimedia.org/wiki/File:Deep_sea_vent_chemistry_diagram.jpg. The right image is redrawn based on discussion with Nick Lane, and using as models an illustration from Woods Hole Oceanographic Institute, and figure 11-03d, Hydrothermal Mounds in "The Origin of Life": https://www.livescience.com/26173-hydrothermal-vent-life-origins.html; "Unicellular Organisms—The Origin of Life": http://www.universe-review.ca/F11-monocell.htm. The latter illustration is modified from Richard Bizley (Science Photo Library) from "The Secret of How Life on Earth Began": http://www.bbc.com/earth/story/20161026-the-secret-of-how-life-on-earth-began.

Figure 18.2: Format based on figure on p. 9: http://www.bio-rad.com/webroot/web/pdf/lsr/literature/Bulletin_5924A.pdf; https://www.difference.wiki/somatic-cells-vs-gametes; figure 2a in D. Duscher et al. (2015), "Stem Cells in Wound Healing: The Future of Regenerative Medicine? A Mini-Review." *Gerontology* 62: 216–25: https://www.difference.wiki/somatic-cells-vs-gametes/.

Figure 18.3: Format based on http://ib.bioninja.com.au/standard-level/topic-3-genetics/33-meiosis/somatic-vs-germline-mutatio.html; https://macscience.wordpress.com/level-2-biology/genetics/somatic-vs-germline-mutations/.

Figure 19.1: Format based on https://thegeneticgenealogist.com/2008/02/15/famous-dna-review-part-iv-jesse-james/.

Figure 20.1: Left and center images based on http://www.dayel.com/blog/2010/10/07/choanoflagellate-illustrations/; right image based on https://www.todaquestao.com/questoes/7675; Amabis e Martho (2001), *Conceitos de biologia* (Sao Paulo: Morderna): http://1.bp.blogspot.com/-fedpfTY6vO8/Tlo—b264-I/AAAAAAAAAws/LJZQcOXaO9M/s1600/14.jpg.

Figure 20.2: Based on http://www.dayel.com/blog/2010/10/07/choanoflagellate-illustrations/.

Figure 22.2: Based on http://www.dayel.com/blog/2010/10/07/choanoflagellate-illustrations/.

Figure 23.2: Format loosely based on Roku Screen Saver: http:// mw40vwind.home/ and image of Early Sea Animals in *Encyclopedia Britannica,* 2015. Details are original. Images used were redrawn using stock images.

Figure 23.5: Format loosely based on Roku Screen Saver: http:// mw40vwind.home/ and on "Early Sea Animals" in *Encyclopedia Britannica,* 2015. Details are original. All images used were redrawn based on stock images.

Figure 24.1: Format of component parts based on https://en.wikipedia.org/wiki/File:Choanoflagellates_(M%C3%A9chnikov).png; https://www.the-scientist.com/the-nutshell/swarm-stimulating-bacterial-enzyme-drives-choanoflagellate-mating-32387; Marina Ruiz Villarreal (LadyofHats); CK-12 Foundation; Creative Commons License. CC BU-NC 3.0 https://www.ck12.org/book/CK-12-Biology/section/18.1/; Ivy Livingstone http://web.augsburg.edu/~capman/bio152/sponges/choanocytes.tiff.jpg.

Figure 24.2: Format of component parts based on https://sites.google.com/site/animalbiologyspring2010/porifera/life-cycle. Art by Mariana Ruiz Villarreal (LadyofHats); CK-12 Foundation; Creative Commons License. CC BU-NC 3.0.

Figure 27.1: Based on G. Jekely (2011), "Origin and Early Evolution of Neural Circuits for the Control of Ciliary Locomotion," *Proceedings of the Royal Society B: Biological Sciences* 278:914–22. PMC3049052.

Figure 29.1: Based on a variety of illustrations found by searching for "acoelomate."

Figure 29.2: Based on images in http://palaeos.com/metazoa/bilateria/bilateria.html.

Figure 30.1: Based on Yassine Mrabet via Wikipedia. Creative Commons Attribution-Share Alike, GNU Free Documentation License.

Figure 31.1: Based on a photograph by Russell Hopcroft published on April 19, 2010, in an article in the *Guardian*: "Microscopic Marine Life": https://bit.ly/2TCd3Yd.

Figure 32.1: Format based on figure 1 in L. Z. Holland et al. (2013), "Evolution of Bilaterian Central Nervous Systems: A Single Origin?" *EvoDevo* 4:27: http://www.evodevojounal.com/content/4/1/27.

Figure 33.1: Public domain image.

Figure 34.2: Based on a variety of illustrations, such as those found in Tetsuto Miyashita (2016), "Fishing for Jaws in Early Vertebrate Evolution: A New Hypothesis of Mandibular Confinement," *Biological Reviews* 91(3):611–57; Chapter 48, "Vertebrates": https://pdfs.semanticscholar.org/66d3/c6327f22f08b1dcd84fb9f8a320610bc7a52.pdf?_ga=2.260516877.1428637398.1553789225-1343959428.1553789225; *Biology Forum,* "The Evolution of the Vertebrate Jaw": https://biology-forums.com/index.php?action=gallery;sa=view;id=101.

Figure 35.1: Based on images on page 26 in N. Shubin (2009), *Your Inner Fish.* (New York: Vintage Books); Wikipedia, "Tiktaalik": https://en.wikipedia.org/wiki/Tiktaalik#/; "Tiktaalik roseae": http://bioweb.uwlax.edu/bio203/f2013/raabe_mic2/.

Figure 38.3: Cortical cell layers from Slideshare: https://bit.ly/2OxdtOw.

Figure 39.1: Reprinted and modified from P. D. MacLean (1949), "Psychosomatic Disease and the 'Visceral Brain': Recent Developments Bearing on the Papez Theory of Emotion," *Psychosomatic Medicine* 11:338–53.

Figure 50.1: Loosely based on http://thebrain.mcgill.ca/flash/a/a_05/a_05_cr/a_05_cr_her/a_05_cr_her_1a.jpg.

Figure 50.2: Location of frontal pole based on figure 2 in M. F. S. Rushworth et al. (2014), "Comparison of Human Ventral Frontal Cortex Areas for Cognitive Control and Language with Areas in Monkey Frontal Cortex," *Neuron* 81:700–713; figures 1 and 2 in Bruno Di Muzio et al.: https://radiopaedia.org/articles/frontal-pole?lang=us.

Figure 50.3: Based on J. K. Rilling et al. (2008), "The Evolution of the Arcuate Fasciculus Revealed with Comparative DTI," *Nature Neuroscience* 11:426–28.

Figure 51.2 Based on "3 Tips To Apply The Cognitive Dissonance Theory In eLearning": https://elearningindustry.com/apply-cognitive-dissonance-theory-elearning.

Figure 54.1: Format based on H. Lau, D. Rosenthal (2011), "Empirical Support for Higher-Order Theories of Conscious Awareness," *Trends in Cognitive Science* 15:365–73.

Figure 55.1: Left: reprinted from J. S. Bruner, A. L. Minturn (1955), "Perceptual Identification and Perceptual Organization," *Journal of General Psychology* 53:21–28. Right: One of many versions of the Dalmatian hidden image: https://www.google.com/search?q=dalmatian+hidden+image&tbm=isch&source=univ&sa=X&ved=2ahUKEwii4sDZkcDhAhUMCuwKHa03CwAQ7Al6BAgHEA0&biw=1493&bih=873#imgrc=-36fZdQC0bNgqM:.

Figure 56.1: Format based on figure 1 in L. R. Squire (2004), "Memory Systems of the Brain: A Brief History and Current Perspective," *Neurobiology of Learning and Memory* 82:171–77.

Figure 56.2: Format based on W. A. Roberts and M. C. Feeney (2009), "The Comparative Study of Mental Time Travel," *Trends in Cognitive Sciences* 13(6), 271–77.

# INDEX

Page numbers in *italics* refer to illustrations and tables.